Mobile Cramming

An FTC Staff Report

July 2014

i

REPORT CONTRIBUTORS

Heather Allen, Attorney, Division of Financial Practices

Duane Pozza, Attorney, Division of Financial Practices

James Trilling, Attorney, Division of Privacy and Identity Protection

Patricia Poss, Attorney, Mobile Technology Unit

Stephanie Rosenthal, Attorney, Division of Financial Practices

Andrew Schlossberg, former Program Specialist, Division of Financial Practices

Patrick Eagan-Van Meter, Program Specialist, Division of Financial Practices

Eric Rodrigues, Paralegal, Division of Financial Practices

TJ Peeler, Visual Information Specialist, Division of Consumer and Business Education

Carrie Gelula, Visual Information Specialist, Division of Consumer and Business Education

Malini Mithal, Assistant Director, Division of Financial Practices

James Reilly Dolan, Associate Director, Division of Financial Practices

Jessica Rich, Director, Bureau of Consumer Protection

Katherine Fallow, Deputy Director, Bureau of Consumer Protection

Katherine Race Brin, Senior Advisor to the Director, Bureau of Consumer Protection

Molly Crawford, Senior Advisor to the Director, Bureau of Consumer Protection

We also thank James Lacko, Deputy Assistant Director, and Devesh Raval, Economist, of the Division of Consumer Protection of the Bureau of Economics, for their contributions to this report.

TABLE OF CONTENTS

EXECUTIVE SUMMARY

The widespread adoption of mobile devices has provided many important benefits to consumers, including the convenience of paying for goods and services using a phone. One mobile payment option is known as "carrier billing" – the ability to charge a good or service directly to a mobile phone account. Industry observers have noted the potential usefulness of carrier billing to consumers, and the industry continues to devote resources to promoting carrier billing as a payment option. As stakeholders have noted, carrier billing of third-party charges may be particularly beneficial for unbanked and underbanked consumers. Additionally, consumers have used text messages to donate funds to a charitable organization, with the charge placed on their mobile phone account.

As carrier billing has developed, however, fraud has become a significant problem for consumers. In particular, mobile cramming – the unlawful practice of placing unauthorized third-party charges on mobile phone accounts – is a significant concern. Mobile cramming occurs when consumers are signed up and billed for third-party services, such as ringtones and recurring text messages containing trivia or horoscopes, without their knowledge or consent. In six recent enforcement actions, the Commission has alleged that such practices have cost consumers many millions of dollars, and in just three of these actions, defendants have agreed to orders imposing judgments totaling more than $160 million.

As part of its consumer protection mission, the Federal Trade Commission ("FTC" or "Commission") works to ensure that consumer protections keep pace with developing technologies and payment mechanisms. In late 2013, following a Commission roundtable on mobile cramming and a number of FTC and state enforcement actions highlighting the prevalence of mobile cramming, the four largest mobile carriers stated that they would discontinue one form of carrier billing for commercial transactions – "Premium SMS" billing. Market participants continue to implement other kinds of carrier billing arrangements, however. In doing so, it is imperative that they keep in mind the same fundamental consumer protection principles that apply regardless of the type of technology used for billing. All stakeholders have an interest in combating cramming on the carrier billing platform, as promoting consumer trust is integral to promoting more widespread adoption of mobile payment systems and further innovation.

Based on evidence reviewed to date, FTC staff recommends certain best practices for industry participants to protect consumers against unwanted charges while enabling innovation and consumer access to another payment mechanism.

First, mobile carriers should give consumers the option to block all third-party charges on their phone accounts. At activation, carriers should inform consumers that third-party charges may be placed on their mobile accounts and carriers should give consumers the opportunity to block all charges at that time. Carriers should also clearly and prominently inform consumers of options to block charges from third parties while accounts are active, including on the carriers' websites. Additionally, carriers should consider offering consumers the ability to block or allow only specific providers, or to block commercial providers only.

Second, advertisements for products or services charged to a mobile bill must not be deceptive. Merchants are in the first instance responsible for ensuring that their practices – including any advertising, marketing, and opt-in processes – are not deceptive. Information about price should be disclosed clearly and conspicuously before charging a consumer's mobile account. Further, carriers should implement reasonable procedures to scrutinize risky or suspicious merchants and terminate or take other appropriate steps against companies engaging in unlawful practices. Such scrutiny should also be applied if a carrier becomes aware that a merchant has run an earlier campaign containing deceptive advertising or engaged in unauthorized billing on landline phones.

Third, it is critical that consumers provide their express, informed consent to charges before they are billed to a mobile account, and that reliable records of such authorizations are maintained. The unreliability of many merchants' claims that they have obtained consumer consent suggests that more centralized control by carriers and intermediaries of the consumer opt-in process and authorization records is needed. Mobile carriers should implement policies, or strengthen existing polices, to investigate and take appropriate action when consumer refund requests, complaints, or other factors indicate that a merchant may be cramming charges without consumers' consent.

Fourth, all charges for third-party services should be clearly and conspicuously disclosed to consumers in a non-deceptive manner. In particular, on a phone bill, the name of the service and any associated bill heading should relate to the product offered and not suggest an affiliation with the carrier's service. Carriers should consider ways to make third-party charges more conspicuous, such as by providing separate subtotals for carrier and third-party charges wherever total charges are disclosed. Carriers also should consider whether consumers who auto-pay their bills would benefit from receiving a separate notification of third-party charges. Further, consumers with prepaid phone plans who do not typically receive phone bills should receive such a notification from the carrier.

Fifth, carriers should implement an effective dispute resolution process. Such a process should be clear and consistent and enable consumers to dispute suspicious charges on their mobile accounts and obtain refunds for unauthorized charges. Consumers have reported difficulties and inconsistent experiences with carriers' dispute resolution policies for third-party charges. Further, given the extensive evidence that consumers are often unaware of third-party charges on their phone bills, carriers should grant consumer refund requests for recurring unauthorized charges that the carrier concludes were crammed, including refunds for the same recurring charge in previous months to the extent it is practicable to identify those prior charges. When a carrier terminates a third party's billing activities due to unauthorized charges, the carrier should notify consumers who incurred charges from the third party to allow them to request a refund.

I.   INTRODUCTION

In recent years, the use of mobile devices has grown so rapidly that mobile devices now outnumber people in the United States, and nearly as many people have a mobile phone as have a bank account.[1] Given these facts, it is no surprise that companies are increasingly offering consumers ways to charge payments for third-party goods and services to their mobile phone accounts – a payment method often referred to as "mobile carrier billing" or "carrier billing." Indeed, carrier billing has been estimated to be the most popular mobile payment system in use in the world today.[2] This payment method offers many potential benefits for consumers who want to use their mobile phones to pay for goods and services and who prefer that the charges be placed on their phone bill rather than on a credit or debit card, for example. In fact, carrier billing may be especially beneficial for unbanked and underbanked consumers.

Carrier billing, however, can also raise concerns that the practice will be misused to charge consumers for items without their consent. This concern has been borne out in consumer complaints and other evidence indicating that in many cases consumers have not authorized charges that have been billed to their phone accounts. Unauthorized third-party charges on mobile phone accounts – a practice known as "mobile cramming" – occurs when consumers are enrolled and billed for third-party services, such as ringtones and recurring text messages containing trivia or horoscopes, without the consumers' knowledge or consent. Third parties either obtain consumers' phone numbers and falsely claim that the consumers have signed up for services and authorized charges, or use deceptive means to obtain consumers' mobile phone numbers – such as by offering free prizes – and then begin charging consumers' phone accounts for recurring third-party charges for purported services unrelated to the offer. These charges often are difficult to locate in phone bills, and many consumers do not notice them or do not understand that such charges are associated with an item unrelated to their phone service. Further, many consumers do not even receive the supposed services for which they are being charged. Mobile cramming thus harms consumers and also undermines the use of carrier billing as a legitimate payment option. As carrier billing is an emerging payment method being actively promoted by carriers, it is in the interest of all stakeholders to take proactive steps now to ensure its reliability and foster consumer trust to enable the payment method to reach its full potential.

For the past two decades, one of the top priorities at the FTC has been ensuring that consumer protections keep pace with emerging technologies, including mobile technologies. Among other things, the FTC has brought enforcement actions to combat mobile cramming and provide restitution to injured

---

[1] *See* Cecilia Kang, *A Nation Outnumbered By Gadgets*, WASHINGTON POST, Oct. 12, 2011 *available at* http://www.washingtonpost.com/business/economy/a-nation-outnumbered-by-gadgets/2011/10/11/gIQAhjdhdL_story.html; BD. OF GOVERNORS OF THE FED. RESERVE, CONSUMER AND MOBILE FINANCIAL SERVICES 2014, 4-5 (2014), *available at* http://www.federalreserve.gov/econresdata/consumers-and-mobile-financial-services-report-201403.pdf [hereinafter "MOBILE FINANCIAL SERVICES REPORT"].

[2] *See* Cary Stemle, *Direct Carrier Billing: The world's most popular mobile payment*, MOBILE PAYMENTS TODAY (Oct. 15, 2013) http://www.mobilepaymentstoday.com/blog/11377/Direct-Carrier-Billing-The-world-s-most-popular-mobile-payment-Infographic.

consumers,[3] recommended the adoption of certain baseline consumer protections,[4] and encouraged public dialogue among industry stakeholders to identify potential solutions. To advance that dialogue, on May 8, 2013, the FTC convened a roundtable of industry participants, consumer advocates, regulators, and other interested parties to discuss and gather information about potential approaches to combat mobile cramming (the "roundtable").[5]

This FTC staff report addresses mobile carrier billing and summarizes information from the roundtable and public comments submitted in connection with the roundtable; data obtained from public sources; and evidence from the FTC's enforcement actions as well as multiple enforcement actions brought by states. It also recommends certain best practices for industry participants to protect consumers against mobile cramming. Specifically, Part II of this report describes mobile third-party billing, including the types of entities involved in the industry. Part III describes mobile cramming and its prevalence. Part IV describes current strategies used to combat mobile cramming and FTC staff's views on whether additional action is needed.

II. MOBILE THIRD-PARTY BILLING

This section describes the potential uses of carrier billing, the types of carrier billing, and the different market participants involved.

A. Uses of Mobile Carrier Billing

As discussed in the FTC's March 2013 staff report on mobile payments and at the roundtable, major phone carriers permit consumers to charge payments for third-party goods and services directly to their mobile phone accounts, which is known generally as "mobile carrier billing" or just "carrier billing,"[6] as an alternative to paying for an item with a credit or debit card, for example.

---

[3] *FTC v. T-Mobile USA, Inc.*, No. 2:14-cv-00967-JLR (W.D. Wash.); *FTC v. Wise Media, LLC*, No. 1:13-cv-1234-WSD (N.D. Ga.); *FTC v. Jesta Digital, LLC*, No. 1:13-cv-01272-JDB (D.D.C.); *FTC v. Tatto, Inc.*, No. 2:13-cv-08912-DSF-FFM (C.D. Cal.); *FTC v. Acquinity Interactive, LLC*, No. 0:14-cv-60166-RNS (S.D. Fla.) (amended complaint filed June 16, 2014); *FTC v. MDK Media, Inc.*, No. 2:14-cv-05099-JFW-SH (C.D. Cal.).

[4] *See* Reply Comment of the Fed. Trade Comm'n, FCC CG Docket No. 11-116 (July 20, 2012), at 7, 12, *available at* http://www.ftc.gov/sites/default/files/documents/advocacy_documents/ftc-reply-comment-federal-communications-commission-concerning-placement-unauthorized-charges/120723crammingcomment.pdf [hereinafter "FTC Reply Comment"].

[5] *See* Press Release, Fed. Trade Comm'n, FTC to Host Mobile Cramming Roundtable May 8 (Mar. 8, 2013), *available at* http://www.ftc.gov/news-events/press-releases/2013/03/ftc-host-mobile-cramming-roundtable-may-8.

[6] *See* FED. TRADE COMM'N STAFF, PAPER, PLASTIC... OR MOBILE? AN FTC WORKSHOP ON MOBILE PAYMENTS 7-8 (2013), *available at* http://www.ftc.gov/sites/default/files/documents/reports/paper-plastic-or-mobile-ftc-workshop-mobile-payments/p0124908_mobile_payments_workshop_report_02-28-13.pdf [hereinafter "FTC MOBILE PAYMENTS REPORT"]. *See also* Sprint Nextel, *Comments for the FTC Mobile Cramming Roundtable* (May 6, 2013), at 1, *available at* http://www.ftc.gov/sites/default/files/documents/public_comments/2013/05/564482-00008-85922.pdf,

Roundtable participants and other industry observers have noted the potential usefulness of mobile carrier billing to consumers. Carrier billing of third-party charges may be useful for consumers who do not have credit cards, or do not want to use them, especially for small transactions. In this way, carrier billing may be beneficial for unbanked and underbanked consumers.[7] One roundtable participant also noted that this payment channel may be an appealing way of transacting for members of the millennial generation, who have grown up in a mobile-centric world and may often use a mobile phone as a primary way to make electronic payments.[8]

In the commercial context, consumers have been able to use carrier billing to pay for such items as text message-based subscription services and digital goods such as virtual currency in a mobile game app. In the charitable context, consumers have used text messages to donate funds to a charitable organization, such as in the aftermath of a natural disaster, with the charge placed on their mobile phone bill.[9] Additionally, consumers have made political contributions by sending text messages and having the charges placed on their mobile phone bills.[10]

[hereinafter "Sprint Nextel Comments"]; Verizon Wireless, *Written Comments for the FTC Mobile Cramming Roundtable* (May 8, 2013), at i, *available at* http://www.ftc.gov/sites/default/files/documents/public_comments/2013/05/564482-00010-85928.pdf, [hereinafter "Verizon Wireless Comments"]; *Manage Mobile Purchases and Subscriptions*, AT&T, http://www.att.com/esupport/article.jsp?sid=52709&cv=820&title=AT#fbid=1T6cx3Wnq50 (last visited July 23, 2014). "Mobile billing" and "carrier billing" refer herein to the general practice of charging a consumer's phone account for third-party services, whether or not the consumer receives a bill. As noted at III.A, *infra*, many consumers, such as consumers with prepaid accounts, do not actually receive a bill.

[7] *See* MOBILE FINANCIAL SERVICES REPORT, *supra* note 1, at 1-2, 5. Beyond the United States, carrier billing may be the only available electronic payment channel for 1.7 billion people in the world who own a mobile phone but do not have a bank account. *See, e.g.*, Oded Israeli, *The "3x factor" of carrier billing in app store purchases*, MOBILE PAYMENTS TODAY (Feb. 1, 2013) http://www.mobilepaymentstoday.com/blog/9789/The-3X-factor-of-carrier-billing-in-app-store-purchases.

[8] *See* Transcript of the Fed. Trade Comm'n Mobile Cramming Roundtable (May 8, 2013), J. Greenwell, BilltoMobile, at 17, *available at* www.ftc.gov/sites/default/files/documents/public_events/Mobile%20Cramming%20Roundtable/30508mob.pdf [hereinafter "FTC Roundtable transcript"]; *see also id.*, M. Niejadlik, Boku, Inc., at 135. References to the FTC Roundtable transcript identify the speaker, the transcript page, and, at the first reference to a particular speaker, the speaker's place of employment.

[9] *See* FTC Roundtable transcript, J. Manis, Mobile Giving Found., at 15-16; *id.*, J. Breyault, National Consumers League, at 31; *id.*, D. Asheim, Give by Cell, at 131-32. *See also* AARON SMITH, PEW INTERNET PROJECT, REAL TIME CHARITABLE GIVING: WHY MOBILE PHONE USERS TEXTED MILLIONS OF DOLLARS IN AID TO HAITI EARTHQUAKE RELIEF AND HOW THEY GOT THEIR FRIENDS TO DO THE SAME (Jan. 12, 2012), *available at* http://www.pewinternet.org/~/media//Files/Reports/2012/Real%20Time%20Charitable%20Giving.pdf.

[10] *See* FTC Roundtable transcript, A. Sege, mQube, at 142. *See also* Derek Johnson, *Political Campaign Text Message Donations – What You Need To Know*, TATANGO (June 11, 2012), http://www.tatango.com/blog/political-campaign-text-message-donations-what-you-need-to-know/.

B. Types of Carrier Billing Arrangements

Historically, third-party charges have often involved a text-messaging component, whereby a consumer purportedly authorizes charges by texting a particular five or six-digit number known as a "short code." Typically, in authorized transactions, the consumer initiates a transaction by sending a text message to a short code or entering his or her phone number on a website. The consumer then receives a text message with additional information about the good or service offered, and the consumer responds to confirm the charge, such as by sending a confirmatory response text. This type of mobile carrier billing is called "Premium SMS" billing, as text messages are also known as "SMS" ("short message service") messages.

Since the adoption of smartphones with advanced mobile web browsing capabilities and the more widespread use of mobile apps, there has been an increasing use of other forms of carrier billing arrangements, known as "direct carrier billing" (or "DCB") arrangements. In DCB arrangements, a consumer does not necessarily need to send or receive a text message to initiate or complete a transaction that is billed to a mobile account. Instead, a consumer can initiate a transaction on a mobile website or within a mobile app, and the merchant can have the charge placed on the consumer's mobile account through back-end arrangements that involve the mobile carriers.

In the past few years, commercial Premium SMS billing has generated billions of dollars of revenue, and one industry participant estimated that third-party charges on mobile bills constituted a $2 to $3 billion dollar annual market.[11] Several roundtable panelists, however, suggested that the Premium SMS market had been declining in size over the last few years, and that one key reason for this is the emergence of mobile apps, which allow consumers to obtain content (such as ringtones) that they may have previously purchased using Premium SMS.[12] The Premium SMS market will likely continue to shrink because, subsequent to the FTC's roundtable, in late 2013, the four largest mobile carriers pledged to discontinue Premium SMS billing for commercial transactions.[13]

In contrast, direct carrier billing, which has generated roughly $300 million in revenue annually, is expected to grow to at least $11 billion worldwide by 2016 for app store purchases alone.[14] DCB arrangements are likely to supplant Premium SMS as the preferred mode of carrier billing, as mobile phone companies have continued to partner with market participants ranging from large technology

---

[11] *See* FTC Roundtable transcript, J. Greenwell, at 23; *see also id.*, J. Breyault, at 14.

[12] *See* FTC Roundtable transcript, J. Manis, at 21; *id.*, J. Greenwell, at 23, 64; *id.*, K. McCabe, Office of Vermont Attorney General, at 54.

[13] Ina Fried, *AT&T, Sprint, T-Mobile, Verizon Dropping Most Premium Text Service Billing in Effort to Combat Fraud*, ALLTHINGSD.COM (Nov. 21, 2013), http://allthingsd.com/20131121/att-sprint-t-mobile-verizon-all-dropping-most-premium-text-service-billing-in-effort-to-combat-fraud/.

[14] *See* FTC Roundtable transcript, J. Greenwell, at 17, 21, 64; *id.*, M. Niejadlik, at 184; Nick Holland & Rich Karpinski, *Carrier Billing: The Latent Operator Opportunity*, YANKEE GROUP (Aug. 21, 2012), *available at* http://www.yankeegroup.com/ResearchDocument.do?id=59249; JORDAN MCKEE, YANKEE GROUP, THE FALL OF PREMIUM SMS PAVES THE WAY FOR DIRECT OPERATOR BILLING DOMINANCE (Feb. 2014), http://bango.com/_/data/support/Fall_of_Premium_SMS.pdf?b_c=YG&b_s=1&b_ct=img.

platforms to smaller mobile payments companies in an effort to make DCB a viable option for mobile payments.[15]

Regardless of the type of carrier billing involved, it is important for companies to keep basic consumer protections in mind, such as providing adequate disclosures containing truthful information and obtaining informed consent for charges. Accordingly, the consumer protection principles embodied in staff's recommendations, discussed in below in Part IV, apply to any Premium SMS arrangements that may still exist or be resumed, DCB arrangements, or other forms of carrier billing that may emerge.

C. Market Participants

The process of placing a charge for third-party services on a phone account typically involves a number of parties, including merchants, billing intermediaries, and mobile carriers. Each of these market participants is under the FTC's jurisdiction when engaged in third-party billing activities.[16]

Merchants provide the goods or services for which consumers are billed on their mobile phone accounts. In Premium SMS arrangements, merchants are often known as "content providers," as many merchants in this area have provided text-based or digital content (such as text-based subscription services or ringtones) to mobile devices. DCB arrangements can be used for many kinds of merchant transactions in which consumers can choose a carrier billing option in lieu of another kind of payment mechanism, like a credit card. Roundtable participants identified, as examples, merchants that use DCB to sell credits on dating, gaming, or social networking sites.[17]

Merchants sign up with **intermediaries** that have contractual agreements with mobile carriers to place charges on mobile phone accounts. In Premium SMS arrangements, these companies often are known as "aggregators."[18] A range of companies acting as billing intermediaries have begun to move

---

[15] *See, e.g.*, FTC Roundtable transcript, J. Greenwell, at 17, 21, 64; *id.*, M. Niejadlik, at 135, 184; Jessie Xu, FACEBOOK, *Helping Monetize the Mobile Web* (June 6, 2012), *available at* https://developers.facebook.com/blog/post/2012/06/06/helping-monetize-the-mobile-web/; *Mobile Operator Billing*, MICROSOFT, http://msdn.microsoft.com/en-us/library/windowsphone/help/jj215902%28v=vs.105%29.aspx (last visited July 23, 2014).

[16] *See* FTC MOBILE PAYMENTS REPORT, *supra* note 6 at 2-3; *FTC v. Verity Int'l, Ltd.*, 443 F.3d 48, 59-60 (2d Cir. 2006); *In re Detariffing of Billing and Collection Servs.*, 102 F.C.C.2d 1150 ¶¶ 30-34 (1986).

[17] *See* FTC Roundtable transcript, J. Greenwell, at 17, 20-21; *id.*, M. Niejadlik, at 135-36.

[18] *All Aggregators*, COMMON SHORT CODE ADMINISTRATION, http://www.usshortcodes.com/partners/find-a-sms-marketing-partner.php#aggregators-tab (last visited July 23, 2014).

into the DCB space.[19] The consumer may or may not directly interact with the billing intermediary at all.

Mobile carriers bill consumers for these transactions, and the carriers get a portion of the consumer payment. The percentage kept by mobile carriers in third-party billing arrangements varies, though in at least one case (*Wise Media*) the carriers received 30-40% of the amount charged to consumers.[20]

A diagram of the billing process is depicted below:

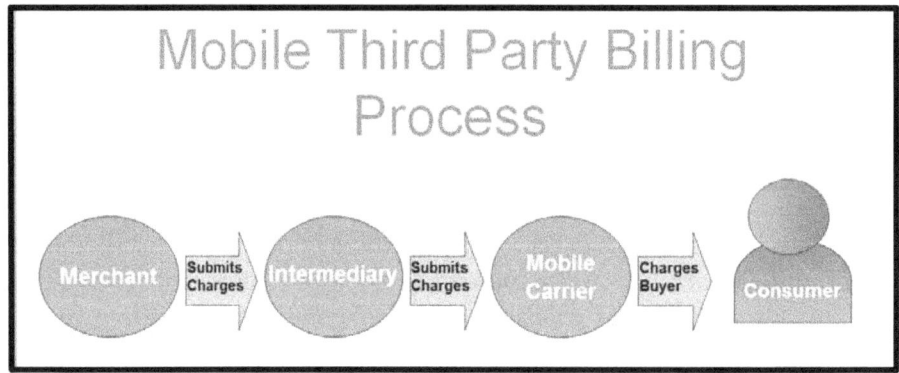

---

[19] *See* FTC Roundtable transcript, J. Greenwell, at 17; *id.*, M. Niejadlik, at 135-136; *id.*, A. Sege, at 143. Companies such as Facebook and Skype have partnered with intermediaries to enable carrier billing in some circumstances on their platforms. *See* Ingrid Lunden, *Facebook Mobile Payment Via Carrier Billing (and Bango) Live In U.S., UK and Germany*, TECHCRUNCH (Sept. 24, 2012), http://techcrunch.com/2012/09/24/facebook-mobile-payments-via-carrier-billing-and-bango-now-live-in-u-s-uk-and-germany/; Ingrid Lunden, *Skype Gets Closer to Mobile Carriers, Inks Deal With Mach For Direct Billing For Skype Credits*, TECHCRUNCH (Sept. 5, 2012), http://techcrunch.com/2012/09/05/skype-gets-closer-to-carriers-inks-deal-with-mach-for-direct-billing/.

[20] *See* Declaration of Andrew R. Schlossberg (Dkt. #3-1), *FTC v. Wise Media, LLC*, No. 1:13-cv-01234-WSD (N.D. Ga. Apr. 16, 2013), at FTC 1 - 000017, ¶ 38 [hereinafter "Wise Media Declaration"]; *see also* Complaint for Permanent Injunction and Other Equitable Relief, *FTC v. T-Mobile USA, Inc.*, No. 2:14-cv-00967 (W.D. Wash. July 1, 2014), ¶ 10 [hereinafter "T-Mobile Complaint"] (alleging that T-Mobile retained typically at least 35% of the charge and in some cases as high as 40%).

III. THE MOBILE CRAMMING PROBLEM

As the FTC, consumer advocates, and other law enforcement and regulatory authorities have observed, mobile cramming has become a significant problem. Part A of this section discusses what mobile cramming is and provides illustrations of how cramming occurs in the mobile context. Part B discusses the evidence demonstrating the extent of mobile cramming. Part C discusses the experiences of other countries in dealing with cramming.

A. What is Cramming?

Cramming is the placement of an unauthorized third-party charge on a phone account.[21] Cramming on landline phone bills has been a problem for many years – likely costing consumers billions of dollars in the last decade.[22] The FTC has brought over thirty cases to stop landline cramming and return money to consumers.[23] As mobile phones have proliferated in recent years, cramming has emerged as a problem in the mobile arena.

Cramming takes advantage of the fact that many consumers do not know that third-party charges can appear on their phone bills; do not notice the charges, which are often buried in the bill under vague terms such as "usage charges"[24] or "monthly service charges" or other terms that suggest a connection to the carrier;[25] or, in the case of consumers with prepaid plans, do not receive a bill at all. Cramming occurs when consumers are signed up and billed for third-party services by merchants, either without any affirmative action by the consumers or after the consumer takes some affirmative act (such as clicking on a mobile webpage or providing a mobile phone number) without understanding that a charge to a mobile phone account will result. Consumers who receive a monthly bill for their mobile accounts

---

[21] This report does not address unauthorized carrier charges on mobile phone bills.

[22] MAJORITY STAFF OF S. COMM. ON COMMERCE, SCI., & TRANSP., OFFICE OF OVERSIGHT & INVESTIGATIONS, UNAUTHORIZED CHARGES ON TELEPHONE BILLS (July 12, 2011), at ii, *available at* http://www.commerce.senate.gov/public/?a=Files.Serve&File_id=3295866e-d4ba-4297-bd26-571665f40756 [hereinafter "S. COMMERCE COMM. CRAMMING REPORT"].

[23] *See, e.g., FTC v. Inc21.com Corp.*, 745 F. Supp. 2d 975 (N.D. Cal. 2010), *aff'd*, 2012 WL 1065543 (9th Cir. Mar. 30, 2012); *FTC v. Hold Billing Servs., Ltd.*, No. 98-cv-00629-FB (W.D. Tex.) (contempt motion filed March 28, 2012); *FTC v. Nationwide Connections, Inc.*, No. 06-80180 (S.D. Fla. Sept. 18, 2008) (stipulated final order); *FTC v. Websource Media, LLC*, No. H-06-1980 (S.D. Tex. July 17, 2007) (stipulated final order); *FTC v. Epixtar Corp.*, No. 03-8511 (S.D.N.Y. Nov. 29, 2006) (stipulated final order); *FTC v. Mercury Mktg. of Del., Inc.*, No. 00-3281, 2004 WL 2677177 (E.D. Pa. Nov. 22, 2004); *FTC v. 800 Connect, Inc.*, No. 03-CIV-60150 (S.D. Fla. Feb. 4, 2003) (stipulated final order); *FTC v. Access Resource Servs., Inc.*, No. 02-CIV-60226 (S.D. Fla. Nov. 4, 2002) (stipulated final order); *FTC v.Cyberspace.com, LLC*, No. C00-1806L, 2002 WL 32060289 (W.D. Wash. July 10, 2002), *aff'd*, 453 F.3d 1196 (9th Cir. 2006).

[24] *See, e.g.*, Wise Media Declaration, *supra* note 20, at Exs. FTC 71, FTC 72, FTC 73.

[25] *See, e.g., id.* at Exs. FTC 66, FTC 78; Exhibits PX 12, PX19 to Exhibits in Support of Plaintiff's *Ex Parte* Application for a Temporary Restraining Order (Dkt. #12), *FTC v. Tatto, Inc.*, No. 2:13-cv-08912-DSF-FFM (C.D. Cal. Dec. 5, 2013).

frequently overlook the charges and pay their phone bills in full.[26] Many consumers may also use auto bill-pay or paperless billing, and/or have family plans that cost a hundred or more dollars each month, making them less likely to notice small charges on their bills.[27] And some prepaid mobile phone consumers do not receive a bill at all – crammed charges are simply deducted from their prepaid balance.[28] As many have pointed out, this ability to charge many consumers without their detecting it makes mobile cramming "almost the perfect scam."[29]

1.   FTC Actions Addressing Mobile Cramming

Three recent FTC cases in which the FTC reached settlements with content providers alleged to be engaged in cramming illustrate several ways that cramming occurs:

- *Wise Media LLC.* The FTC filed suit in April 2013 against merchant Wise Media, LLC, which purported to sell recurring subscriptions to text message services providing "love tips," horoscopes, diet tips, and similar kinds of "alerts" for $9.99 a month.[30] The company claimed that consumers signed up for the services by entering their information into websites, receiving PIN codes by text messages, and inputting the PINs into the websites. Consumers who discovered the charges, however, widely reported that they had never heard of Wise Media or signed up for the services, suggesting they were simply billed without authorization.[31] In November 2013, a court entered a stipulated order with a judgment for

---

[26] *See, e.g., Inc21.com*, 745 F. Supp. 2d at 1001; Memorandum in Support of Motion for Temporary Restraining Order, at 10-11, *FTC v. Wise Media, LLC*, No. 1:13-cv-1234-WSD (N.D. Ga. Apr. 16, 2013) [hereinafter "Wise Media TRO Memo"]; FTC Roundtable transcript, A. Sege, at 145; *id.*, J. Chilsen, Citizens Utility Board, at 73; *id.*, C. Witteman, California Public Utilities Commission, at 100; *id.*, P. Singer, Office of Texas Attorney General, at 102.

[27] *See* FTC Roundtable transcript, J. Breyault, at 31-32; Brad Tuttle, *Groceries or Mobile Phone? Plenty of Consumers Spend More on the Latter*, TIME MAGAZINE, Sept. 13, 2012, *available at* http://business.time.com/2012/09/13/groceries-or-mobile-phone-plenty-of-consumers-spend-more-on-the-latter/ (noting that nearly half of Americans with mobile phones said their monthly bill comes to $100 or more per month); Anton Troianovski, *Cellphones Are Eating the Family Budget*, WALL ST. J., Sept. 28, 2012, *available at* http://online.wsj.com/article/SB10000872396390444083304578018731890309450.html.

[28] *See, e.g., Cricket Wireless Terms and Conditions of Service*, CRICKET WIRELESS, https://www.cricketwireless.com/terms (last visited July 23, 2014); *Choose a Prepaid Plan. Pick Your Device, View Plan Details*, VERIZON WIRELESS, http://www.verizonwireless.com/b2c/prepay/processPrePayRequest.do?type=ppmonthBASIC (last visited July 23, 2014).

[29] *See, e.g.*, FTC Roundtable transcript, J. Breyault, at 32; *see also id.*, J. Chilsen, at 72-73.

[30] Complaint for Permanent Injunction and Other Equitable Relief, at 7-8, *FTC v. Wise Media, LLC*, No. 1:13-cv-1234-WSD (N.D. Ga. Apr. 16, 2013), *available at* http://www.ftc.gov/sites/default/files/documents/cases/2013/04/130417wisemediacmpt.pdf.

[31] *See* Wise Media TRO Memo, *supra* note 26, at 6-9.

more than $10 million and a ban that prohibits Wise Media from placing charges on mobile phone bills altogether.[32]

- *Jesta Digital, LLC.* In this case, filed in August 2013, the FTC alleged that the defendant lured consumers into purchasing a monthly subscription for ringtones using deceptive practices.[33] According to the complaint allegations, some consumers saw banner ads on their mobile devices that falsely claimed a virus had been detected. Clicking on the ad led to a screen with a button stating "Get Now" above the phrase "Protect your Android [phone] today." Consumers who clicked "Get Now," and then a button on a subsequent page marked "Subscribe," were then subscribed to the $9.99 per month ringtone subscription plan, though the nature and cost of the subscription were never adequately disclosed. Indeed, some consumers were subscribed even if they clicked on parts of the screen other than the "subscribe" button. Moreover, if consumers actually attempted to subscribe and download Jesta's so-called anti-virus software to their mobile devices, the download often failed. Jesta used a process known as WAP or Wireless Access Protocol billing,[34] which captures consumers' phone numbers from a mobile device, to obtain consumers' purported authorization for the charges. Thus, consumers never even entered their phone number prior to being billed.[35]

- *Tatto, Inc. & Bullroarer, Inc.* In a third case, filed in December 2013, the FTC alleged that another widespread mobile cramming operation engaged in similar deceptive practices. For example, the FTC alleged that the defendants ran websites that promised consumers offers such as free Justin Bieber tickets.[36] When consumers attempted to claim these offers, they were asked for a mobile phone number. After following the instructions provided, consumers did not receive the Justin Bieber tickets, yet, the Commission has alleged, it is

---

[32] Stipulated Order for Permanent Injunction and Monetary Judgment Against Defendants Brian M. Buckley and Wise Media, LLC, at 4-6, *FTC v. Wise Media, LLC*, No. 1:13-cv-1234-WSD (N.D. Ga. Nov. 22, 2013), *available at* http://www.ftc.gov/sites/default/files/documents/cases/131121wisemediabuckleystip.pdf [hereinafter "Wise Media Settlement"].

[33] Complaint for Permanent Injunction and Other Equitable Relief, at ¶¶ 8-25, *FTC v. Jesta Digital, LLC*, No. 1:13-cv-01272 (D.D.C. Aug. 20, 2013), *available at* http://www.ftc.gov/sites/default/files/documents/cases/2013/08/130821jestacmpt.pdf [hereinafter "Jesta Digital Complaint"].

[34] As discussed in note 128, *infra*, WAP opt-in involves consumers responding to an offer displayed on the mobile web by clicking on a confirmation button from the phone two separate times. This process captures the consumer's phone number without the need for the consumer to enter it manually.

[35] *See* Jesta Digital Complaint, *supra* note 33, at ¶ 24.

[36] Complaint for Permanent Injunction and Other Equitable Relief, at 9, *FTC v. Tatto, Inc.*, No. 2:13-cv-08912-DSF-FFM (C.D. Cal. Dec. 5, 2013), *available at* http://www.ftc.gov/sites/default/files/documents/cases/131216bullroarercmpt.pdf.

likely that consumers were instead signed up for the defendants' subscription plans.[37] The primary corporate defendants have agreed to a partially suspended judgment of $150,153,283.[38]

The Commission has also pursued two recent additional actions against content providers that raise similar issues. In one case, the Commission has alleged that certain defendants sent text messages promising free $1,000 gift cards and iPads as a way to deceive consumers into "confirming" their phone number and entering a PIN on a website, which resulted in consumers being signed up for unwanted premium text messaging services and incurring charges of $9.99 per month on their mobile phone accounts.[39] In another case, the Commission has alleged that a content provider similarly used the lure of "free" gift cards to collect consumers' phone numbers and crammed consumers for subscription services such as horoscope alerts.[40]

Additionally, the Commission recently filed suit in federal district court against T-Mobile USA, alleging that T-Mobile charged consumers for these kinds of monthly text message subscriptions purportedly offered by third-party merchants that, in many cases, were not authorized by consumers. The complaint alleges that T-Mobile deceptively described these charges on its phone bills in a manner that made it difficult for consumers to uncover them. For example, for consumers who reviewed an online summary of their bills, the complaint alleges the third-party charges were lumped together in a line item labeled "Use Charges" that could include charges for both T-Mobile's own services, such as for text messages, and for third-party charges. Additionally, according to the complaint, T-Mobile continued to charge consumers even after becoming aware of telltale signs that the charges were unauthorized. The complaint alleges that T-Mobile's internal documents showed that consumers were increasingly calling T-Mobile to complain about unauthorized third-party charges; that large numbers of consumers sought refunds and the refund rate for some subscriptions was higher than 40% in some months; and that T-Mobile continued to charge consumers for third-party merchants for years after those merchants were the subject of law enforcement or other legal action for cramming,

---

[37] *See id.*; Memorandum In Support of Plaintiff's Ex Parte Application For Temporary Restraining Order With An Asset Freeze and Other Equitable Relief, And Order to Show Cause Why A Preliminary Injunction Should Not Issue, at 12, *FTC v. Tatto, Inc.*, No. 2:13-cv-08912-DSF-FFM (C.D. Cal. Dec. 5, 2013) [hereinafter "Tatto TRO Memo"].

[38] *See* Stipulated Order for Permanent Injunction and Monetary Judgment Against Defendants Tatto, Inc., Shaboom Media, LLC, Bune, LLC, Mobile Media Products, LLC, Chairman Ventures, LLC, Galactic Media, LLC, Virtus Media, LLC, and Lin Miao, *FTC v. Tatto, Inc.*, No. 2:13-cv-08912-DSF-FFM (C.D. Cal. June 11, 2014), *available at* http://www.ftc.gov/system/files/documents/cases/140613bullroarerstiporder.pdf. The judgment was partially suspended based on defendants' ability to pay, but the defendants that have settled to date have surrendered more than $10 million in assets to be used for consumer redress.

[39] Amended Complaint for Permanent Injunction and Other Equitable Relief, *FTC v. Acquinity Interactive, LLC*, No. 14-60166-CIV (S.D. Fla. June 16, 2014), *available at* http://www.ftc.gov/system/files/documents/cases/140707revenuepathcmpt.pdf.

[40] Complaint for Permanent Injunction and Other Equitable Relief, *FTC v. MDK Media, Inc.*, No. 2:14-cv-05099-JFW-SH (C.D. Cal. July 3, 2014).

news articles detailing cramming behavior, and industry auditor alerts detailing deceptive practices in which those merchants were engaged.[41]

2. Other Federal and State Initiatives Addressing Mobile Cramming

The FTC is joined by a number of partners in its efforts to protect against cramming. On the federal level, for several years the Federal Communications Commission ("FCC") has been engaged in a rulemaking proceeding in which it is proposing additional truth-in-billing rules to help consumers detect crammed charges on phone bills, and is considering whether to extend those rules to mobile bills.[42] The FTC has filed comments in that proceeding calling for stronger consumer protections.[43] And the FCC recently announced it launched an investigation into T-Mobile with regard to cramming.[44]

Congress also has focused on the problem of mobile cramming. The Chairman of the Senate Committee on Commerce, Science, and Transportation has issued requests to mobile carriers and billing aggregators to provide information on cramming in order for the Committee to assess potential legislative solutions.[45] Chairman Rockefeller has noted that, based on review of public complaints by his staff, "[c]onsumers continue to complain that they are experiencing unauthorized charges on their wireless bills for 'services' they did not order and do not use," and "[t]he types of so-called 'services' that are appearing on these consumers' wireless bills . . . are alarmingly similar to the services shown to be fraudulent on wireline telephone bills."[46]

---

[41] *See* T-Mobile Complaint, *supra* note 20, at ¶¶ 11-36.

[42] Public Notice, Fed. Commc'ns Comm'n, Consumer and Governmental Affairs Bureau Seeks to Refresh the Record Regarding "Cramming" (Aug. 27, 2013), *available at* http://transition.fcc.gov/Daily_Releases/Daily_Business/2013/db0827/DA-13-1807A1.pdf [hereinafter "FCC Refresh the Record"].

[43] *See* FTC Reply Comment, *supra* note 4, at 1, 5-7, 12.

[44] Press Release, Fed. Commc'ns Comm'n, FCC Investigates Cramming Complaints Against T-Mobile (July 1, 2014), *available at* http://www.fcc.gov/document/fcc-investigates-cramming-complaints-against-t-mobile.

[45] *See* Press Release, S. Comm. on Commerce, Science, and Transp., Rockefeller Vows to Avert Wireless Cramming Scams on Consumers (Mar. 1, 2013), *available at* http://www.commerce.senate.gov/public/index.cfm?p=HearingsandPressReleases&ContentRecord_id=cd0edc13-b355-4d4e-9619-7035329daa1a&ContentType_id=77eb43da-aa94-497d-a73f-5c951ff72372&Group_id=165806cd-d931-4605-aa86-7fafc5fd3536&MonthDisplay=3&YearDisplay=2013; Press Release, S. Comm. on Commerce, Science, and Transp., Rockefeller Questions Billing Aggregators on Wireless Cramming (Mar. 22, 2013), *available at* http://www.commerce.senate.gov/public/index.cfm?p=HearingsandPressReleases&ContentRecord_id=07f5f79c-f6c6-4ca7-a03b-ca76ac5d3962&ContentType_id=77eb43da-aa94-497d-a73f-5c951ff72372&Group_id=165806cd-d931-4605-aa86-7fafc5fd3536&MonthDisplay=3&YearDisplay=2013.

[46] *See, e.g.*, Letter from Sen. John D. Rockefeller IV to Randall Stephenson, March 1, 2013, *available at* http://www.commerce.senate.gov/public/?a=Files.Serve&File_id=2610b9ba-a3d1-43eb-a94f-d7505f354680.

The states have been active in this area as well. Based on their experience reviewing complaints and conducting investigations and enforcement actions, the Attorneys General of 36 States, the District of Columbia, Guam, Puerto Rico, and the Virgin Islands submitted a comment to the FTC, in connection with the roundtable, urging action to address mobile cramming. The comment notes that they "continue to receive complaints from consumers that charges . . . appeared out of the blue on their phone bills without their authorization and for goods and services that the consumers neither requested nor used,"[47] pointing out that the complaints share common themes "consistent across the country and across time."[48] Given the scope of fraudulent conduct, the Attorneys General have noted that they are "concerned that, under this system, too much responsibility has been placed in the hands" of third parties to obtain consumers' consent – "parties for whom there is a significant incentive to bill for services whether or not they have obtained authorization from consumers."[49]

Indeed, multiple state enforcement authorities have been active in combating mobile cramming. States, including Florida,[50] New York,[51] Texas,[52] and Washington,[53] have targeted deceptive practices used to sign up consumers unknowingly for unwanted subscription services. For example, as discussed below in at Part IV.B.1.a, the State of Texas recently sued a mobile cramming operation that allegedly used deceptive means on websites and in text messages to sign up consumers for services without consent, and it has also filed suit against a Premium SMS aggregator, which it alleged participated in merchants' deceptive practices to enroll consumers in the merchants' programs and assisted them in avoiding detection by consumers.[54]

---

[47] *See* Nat'l Ass'n of Att'y Gen., *Comments for the FTC Mobile Cramming Roundtable* (June 24, 2013), at 1, *available at* http://www.ftc.gov/sites/default/files/documents/public_comments/2013/06/564482-00015-86106.pdf [hereinafter "NAAG Comments"].

[48] *See id.* at 3-4. The Attorneys General also note that some consumers who received text messages suggesting they had been opted in to a service responded "STOP" but were still signed up. *See id.* at 11. Many of these themes mirror landline cramming complaints.

[49] *Id.* at 12.

[50] *See id.* at 7-8. *See also, e.g.*, Assurance of Voluntary Compliance, *In re Verizon Wireless Services LLC & Alltel Commc'ns, LLC*, Case Nos. L08-3-1035 & L08-3-1034 (Fla. Att'y Gen. June 16, 2009), *available at* http://myfloridalegal.com/webfiles.nsf/WF/KGRG-7TAJQ2/$file/VerizonAVC.pdf.

[51] NAAG Comments, *supra* note 47, at 8.

[52] *See* Plaintiff's Original Petition, at ¶¶ 18-24, *State of Texas v. Cellzum.com, LLC*, No. D-1-GV-13-000629 (Travis County, Tex. Dist. Ct. July 11, 2013), *available at* https://www.oag.state.tx.us/newspubs/releases/2013/Cellzum_POP_070813.pdf.

[53] NAAG Comments, *supra* note 47, at 9.

[54] *See* Plaintiff's Original Petition, *State of Texas v. Mobile Messenger U.S. Inc.*, No. D-1-GV-13-001256 (Travis County, Tex. Dist. Ct. Nov. 6, 2013), *available at* https://www.oag.state.tx.us/newspubs/releases/2013/Mobile-Messenger-POP.pdf [hereinafter "Mobile Messenger Complaint"].

B. Prevalence of Mobile Cramming

One of the key issues discussed by roundtable participants was the prevalence of mobile cramming. Publicly available information, including evidence about the rates at which consumers obtain refunds for mobile third-party charges and the volume of those refunds, indicates that cramming is a significant problem.[55] This section discusses that evidence, as well as indications that these consumer complaints understate the full extent of actual consumer injury.

1. Carrier Refund Rates

Carrier refund rates for commercial third-party billing – the ratio of a carrier's refunds to charges billed for a particular period of time, such as a month – have been significant. Carriers keep track of these refund rates and sometimes keep a greater share of the revenue from third-party charges for merchants that have high refund rates or a high number of consumer complaints than they do for those with lower refund rates or complaints.[56] Between January 2011 and September 2012, mobile carriers reported data to the California Public Utilities Commission (CPUC) showing that the overall refund rate for third-party charges for California mobile consumers was approximately 12-13% each month, with a high of 18.6% in one month.[57] Wise Media, one of the companies the FTC sued for mobile cramming, had refund rates exceeding 40% in some months for certain services.[58]

High refund rates are often considered to be indicia of fraudulent conduct, and the refund rates for commercial carrier billing appear to be an order of magnitude higher than refund rates for other types of billing. For example, charitable donations charged to a mobile bill and processed through the Mobile Giving Foundation typically have a refund rate of under 1% overall.[59] In the credit card industry, public evidence from one payment processing network shows that the average credit card chargeback rate is around 0.2%, and that a chargeback rate of 1% for any one merchant in a month triggers an

---

[55] Unless otherwise noted, the following discussion refers to cramming by commercial merchants. The FTC has not seen evidence to date of cramming in the context of charitable or political donations.

[56] *See* FTC Roundtable transcript, M. Altschul, CTIA – The Wireless Association, at 24-25; Comments of CTIA – The Wireless Association, FCC CG Docket No. 11-116 (June 25, 2012), at 5-6; Comments of Sprint Nextel Corporation, FCC CG Docket No. 11-116 (Oct. 24, 2011), at 7.

[57] *See* FTC Roundtable transcript, C. Witteman, at 91; Cal. Pub. Util. Comm'n, *Letter to Melanie K. Tiano, S. Comm. on Commerce, Science, and Transp., Re: Inquiry Regarding Wireless California Cramming Complaint Data* (Jan. 31, 2013), *available at* http://apps.fcc.gov/ecfs/document/view;jsessionid=ng1nRV9Wynn82QmDpwMwChl87kFMQhvhcqH5Q21ZLqg b1cJN2Tn1!-1705390101!956499833?id=7022119377 [hereinafter "CPUC Letter"] (percentages extracted from data in the letter). California regulations require wireless carriers to submit monthly data on third-party charges and refunds. *See* CAL. PUB. UTIL., *Final Decision Adopting California Telephone Corporation Billing Rules*, Rulemaking No. 00-02-004 (Nov. 3, 2010), Attach. A at 8 [hereinafter "CPUC Rule"].

[58] *See* Wise Media TRO Memo, *supra* note 26, at 10; Wise Media Declaration, *supra* note 20, at FTC 1 – 000012, ¶ 25, Ex. 21.

[59] *See* FTC Roundtable transcript, J. Manis, at 58.

investigation.[60] Similarly, for ACH (Automated Clearing House) transactions, a 1% rate of chargebacks for unauthorized ACH transactions by a merchant may trigger an enforcement proceeding.[61]

Carriers sometimes set threshold refund rates for disciplining merchants, such as prohibiting them from placing charges on consumers' bills.[62] One carrier, for example, has terminated Premium SMS merchants with an 8% or higher refund rate.[63] Nevertheless, at least one carrier continued billing consumers for Wise Media charges even after the company's monthly refund rate for some services exceeded 40%.[64]

As high as the carrier billing refund rates are, available evidence indicates that they understate the extent of refund requests from consumers. On the one hand, industry representatives stated at the roundtable that carriers have liberal refund policies and that, in some cases, a carrier may provide a refund even if someone else in the complaining consumer's household authorized the charge, or when the complaint is about an issue other than cramming.[65] Consumers, however, have often reported difficulties requesting refunds for crammed charges from carriers. Many complain that carriers refuse to give more than two months' worth of refunds, even if consumers learn that crammed charges have appeared on their bills for longer periods of time.[66] In other instances, carriers have told consumers to contact the merchant for a refund, a request that, even if the consumer is able to make contact with the merchant, the merchant often denies.[67]

2. Complaint Information

Government agencies also have received a significant number of complaints related to third-party charges on mobile accounts. For example, the FTC's Consumer Sentinel database shows that

---

[60] *See FTC v. Commerce Planet, Inc.*, 878 F. Supp. 2d 1048, 1075 (C.D. Cal. 2012); *FTC v. Grant Connect, LLC*, 827 F. Supp. 2d 1199, 1222 (D. Nev. 2011).

[61] *See* NACHA – The Electronic Payments Association, NACHA Operating Rules, Art. II, § 2.17.2 (ODFI Return Rate Reporting) and Appendix 10 §§ 10.2.2, 10.4.3 (Initiation of a Rules Enforcement Proceeding) (2014) [hereinafter "NACHA Operating Rules"]. The 2014 edition of the NACHA Rules is available at www.achrulesonline.org.

[62] *See* FTC Roundtable transcript, J. Bruner, Aegis Mobile, at 82; *id.*, M. Altschul, at 146.

[63] *See* Wise Media Declaration, *supra* note 20, at FTC 1 - 000012-13, ¶ 27.

[64] *See id.* at Ex. FTC 21.

[65] *See* FTC Roundtable transcript, M. Altschul, at 146, 152-53; *id.*, M. Niejadlik, at 147-48; *id.*, A. Sege at 149-150; Sprint Nextel Comments, *supra* note 6, at 2, 6; Verizon Wireless Comments, *supra* note 6, at 5.

[66] *See* FTC Roundtable transcript, P. Singer, at 102; Wise Media Declaration, *supra* note 20, at Ex. FTC 69, FTC; JANE KOLODINSKY, CTR. FOR RURAL STUDIES, UNIV. OF VT., MOBILE PHONE THIRD-PARTY CHARGE AUTHORIZATION STUDY (2013), Appendix C at 6-8, *available at* http://www.atg.state.vt.us/assets/files/Mobile%20Phone%20Third-Party%20Charge%20Authorization%20Study.pdf; NAAG Comments, *supra* note 47, at 3-4.

[67] *See* Wise Media TRO Memo, *supra* note 26, at 11-12; FTC Roundtable transcript, J. Breyault, at 55; *id.*, P. Singer, at 102; NAAG Comments, *supra* note 47, at 4.

consumers have reported over 1800 complaints of unauthorized charges on wireless bills since 2010.[68] As noted in greater detail below, while this is a significant number, the number of cramming complaints to the FTC likely significantly understates the actual extent of consumer injury. For example, the number is dwarfed by the number of refund requests to carriers.[69] In the *Wise Media* case alone, evidence obtained in the FTC's investigation showed that carriers granted over 190,000 refunds to consumers for Wise Media's charges.[70] And in the *Tatto* case, the FTC's evidence showed that the carriers granted complaining consumers more than 1.2 million refunds for charges by the group of companies the FTC sued.[71] Thus, while the FTC's complaint database showed a total of approximately 1800 complaints, carriers made almost 1.4 million refunds to complaining consumers in two cases alone.

Similarly, the CPUC reports that it has never received more than 10 cramming complaints in a month during the 2011-2012 time period for which it provided data, which is a tiny fraction of the carrier refunds provided during that time. Specifically, for all of 2011 and much of 2012, the CPUC reports that carriers have provided no fewer than 160,000 refunds a month to California consumers and have sometimes provided over 300,000 refunds a month. In every month for which the total number of refunds was reported, carriers provided at least 20,000 times more refunds than complaints received by the CPUC, meaning that complaints reported to the state agency represented less than 0.005% of the refunds actually given in a month.[72]

3. Other Efforts to Estimate the Extent of Cramming

Other stakeholders have also sought to estimate the extent to which third-party charges are crammed as a percentage of the overall carrier billing market. At the roundtable, the Vermont Attorney General's office presented the results of a study addressing the size and nature of the mobile cramming problem in Vermont. The study found that, out of a random sample of Vermont consumers who had incurred third-party charges on their phone bills, 60% of consumers reported that neither they nor anyone in their household had authorized the third-party charges listed on their bills. Over 55% said

---

[68] *See* FTC Reply Comment, *supra* note 4, at 5. These complaints are consistent in their content with hundreds if not thousands of consumer complaints in online message boards. *See*, for example, the many reports of cramming at www.smswatchdog.com, www.textcomplaints.com, and the forums on the carriers' websites. According to the National Association of Attorneys General, several State Attorneys General recently reviewed cramming complaints received by 28 states and found over 750 mobile cramming complaints, predominantly from the last few years. *See* NAAG Comments, *supra* note 47, at 2.

[69] The evidence suggests that consumers who become aware of crammed charges and take the time to complain are far more likely to contact their carrier than government agencies.

[70] Wise Media TRO Memo, *supra* note 26, at 10.

[71] Tatto TRO Memo, *supra* note 37, at 10.

[72] CPUC Letter, *supra* note 57, at 2; *see also* FTC Roundtable transcript, C. Witteman, at 91. The State of Vermont also reports that it has received around two dozen mobile cramming complaints in the last seven years, but it identified hundreds of consumers whose responses indicated that they had been crammed when it proactively surveyed consumers who had incurred third-party charges on their mobile phone bills. FTC Roundtable transcript, K. McCabe, at 12.

they were unaware of the charges until the survey.[73] The Citizens Utility Board in Illinois has also announced that an analysis of mobile phone lines by an outside research firm found that 44% of all third-party charges appeared to be fraudulent.[74]

On the other hand, CTIA has criticized the methodology of the Vermont survey and submitted the testimony of an expert who analyzed the Vermont study on behalf of the four major mobile phone carriers. According to CTIA, the expert "found that the survey methodology underlying the Vermont Study did not comply with core principles of objective research, and further concluded that 'the Vermont Study is neither a valid nor reliable measure of the extent to which, if any, Vermont mobile phone users have problems with unauthorized third-party charges on their bills.'"[75]

Although no industry participant attempted to estimate the extent of cramming at the roundtable or in a public comment, industry members also have questioned the prevalence of mobile cramming. CTIA has stated that mobile cramming is "not a significant consumer concern," pointing to the number of mobile cramming complaints reported by federal agencies and the industry's voluntary actions to address cramming.[76] Mobile phone carriers, including Verizon, AT&T, and Sprint, have publicly explained steps they have taken to prevent cramming on the mobile platform and suggest that industry standards are working to prevent fraud in this space.[77] In addition, a representative from the Mobile Giving Foundation, who previously served as chairman of the Mobile Marketing Association, suggested at the roundtable that mobile cramming is on the decline given steps taken by the carriers.[78]

Despite extensive evidence of consumers complaining to and seeking refunds for crammed charges, much of it developed through enforcement actions, industry representatives have not attempted to estimate the full extent of crammed charges based on their own data. Instead, industry representatives primarily rely on data about public complaints to government agencies, which, as noted above, represent a small fraction of actual cramming.

---

[73] See KOLODINSKY, supra note 66, at 6-7; FTC Roundtable Transcript, K. McCabe, at 12, 28-29.

[74] See Citizens Utility Board, Analysis: Frequency Of Cellphone 'Cramming' Scam Doubles In Illinois, CUB Concerned Wireless Customers Targeted As Landline Laws Tighten (Dec. 4, 2012), available at http://www.citizensutilityboard.org/newsReleases20121204_CellphoneCramming.html.

[75] CTIA – The Wireless Ass'n, Re: "Mobile Cramming" Roundtable, Project No. P134830, (June 24, 2013), at 1, available at http://www.ftc.gov/policy/public-comments/comment-564482-00016. Even if the percentage of charges that were unauthorized was somewhat lower than those reported in the Vermont study, this would still point to a substantial volume of unauthorized third-party billing.

[76] Comments of CTIA – The Wireless Ass'n, FCC CG Docket No. 11-116 (June 25, 2012), at 3-6.

[77] See Verizon Wireless Comments, supra note 6; Sprint Nextel Comments, supra note 6. Specific industry approaches are discussed in detail in Part IV, infra.

[78] See FTC Roundtable transcript, J. Manis, at 15.

4.   Unreported Consumer Injury

The FTC's enforcement actions and other experience further illustrate the fact that reported consumer complaints and refund requests do not reflect the full extent of injured consumers. This is sometimes described as the level of complaints representing the "tip of the iceberg" of actual consumer injury.[79]

- At the time of filing suit in *FTC v. Inc21.com Corp.*, a landline cramming case, the FTC was able to point to about 280 formal consumer complaints to government agencies.[80] An expert survey conducted during the course of litigation and admitted by the court showed that nearly 97% of billed consumers did not authorize the charges.[81] After prevailing in litigation, the FTC provided monetary redress to over 139,357 consumers.[82]

- In *FTC v. Wise Media, LLC*, the FTC noted at the time of filing that it had received over 100 Consumer Sentinel complaints regarding Wise Media's monthly charges of $9.99, but the defendants later agreed to a stipulated judgment amount of over $10 million.[83]

- A 2004 study by the FTC's Bureau of Economics and Bureau of Consumer Protection found that only 8.4% of defrauded consumers complained to a government agency or the Better Business Bureau, meaning that the vast majority of fraudulent conduct is not officially reported to a government agency.[84]

One reason that refund requests and complaints represent only a small fraction of all crammed charges is that many consumers do not notice charges on their phone bills, in part because the descriptions of the charges are often buried and uninformative. As a result, these consumers never submit a complaint or seek a refund at all. In the context of landline cramming, a court-accepted survey in the *Inc21.com* case found that only five percent of consumers were even aware of the unauthorized charges from review of their bills.[85] Several panelists at the roundtable also commented that consumers may not know that their bills can be charged for third-party services, and that many of these small

---

[79] *See, e.g.*, FTC Reply Comment, *supra* note 4, at 6-7; Consumers Union, *Comments for the FTC Mobile Cramming Roundtable* (June 24, 2013), at 2-3, *available at* http://www.ftc.gov/sites/default/files/documents/public_comments/2013/06/564482-00014-86099.pdf.

[80] *See* Exhibits in Support of Motion for Temporary Restraining Order, *FTC v. Inc21.com Corp.*, No. 3:10-cv-00022-WHA (N.D. Cal. Jan. 5, 2010), at Ex. 1, at 4.

[81] *See FTC v. Inc21.com Corp.*, 745 F. Supp. 2d at 1001.

[82] *See* Press Release, Fed. Trade Comm'n, FTC Returns More Than $5.4 Million to Victims of Massive Cramming Scam (Sept. 30, 2013), *available at* http://www.ftc.gov/opa/2013/09/inc21.shtm.

[83] *See* Wise Media Declaration, *supra* note 20, at FTC 1 - 000013-14, ¶ 30; Wise Media Settlement, *supra* note 32, at 6. The judgment was partially suspended based on defendants' ability to pay.

[84] FED. TRADE COMM'N STAFF, CONSUMER FRAUD IN THE UNITED STATES: AN FTC SURVEY (2004), *available at* http://www.ftc.gov/sites/default/files/documents/reports/consumer-fraud-united-states-ftc-survey/040805confraudrpt.pdf.

[85] *See Inc21.com Corp.*, 745 F. Supp. 2d at 1001.

charges can be labeled deceptively.[86] In the *Wise Media* case, for example, consumers complained that the charges were buried in lengthy phone bills (for example, a $9.99 charge was listed on page 18 of one consumer's bill).[87] In another case, the allegedly crammed charge was listed on page 123 of a consumer's online bill.[88] Third-party charges have often been placed in the bill under vague terms such as "usage charges" or "monthly service charges" or other terms that suggest a connection to the carrier.[89] Additionally, some consumers, such as those with prepaid mobile phone plans, may not receive a bill at all.[90]

C. International Views

Consumer protection authorities outside the United States have sought to combat unauthorized third-party charges on consumers' mobile phone bills as well. For instance, Canada's Competition Bureau filed a court action in fall 2012 against three of the major Canadian mobile phone carriers for misleading conduct related to Premium SMS charges.[91] In the United Kingdom, regulators recently have seen a surge in cramming complaints and thus have initiated a series of investigations and have imposed a number of fines.[92] Consumers in other countries, like Australia, Russia, and China, have experienced similar problems.[93] The London Action Plan, an international anti-spam network consisting of 27 countries around the world, released a report in October 2012 called "Best Practices to

---

[86] *See* FTC Roundtable transcript, A. Sege, at 145; *id.*, C. Witteman, at 100; *id.*, P. Singer at 86-87; *see also* NAAG Comments, *supra* note 47, at 3, 6.

[87] *See* Wise Media TRO Memo, *supra* note 26, at 6.

[88] *See* Graphic, "Excerpts from an actual T-Mobile bill," *available at* http://www.ftc.gov/system/files/attachments/press-releases/ftc-alleges-t-mobile-crammed-bogus-charges-customers-phone-bills/tmobile-samplebill.pdf.

[89] *Supra* Part III.A, at 10.

[90] *Id.*

[91] *See* FTC Roundtable transcript, L. Bryenton, Competition Bureau, Canada, at 49. *See also* Press Release, Competition Bureau (Canada), Competition Bureau Sues Bell, Rogers and Telus for Misleading Consumers: Bureau Seeks Customer Refunds and $31 Million in Penalties (Sept. 14, 2012), *available at* http://www.competitionbureau.gc.ca/eic/site/cb-bc.nsf/eng/03498.html.

[92] *See* PhonepayPlus, *Comments for the FTC Mobile Cramming Roundtable* (July 16, 2013), at 2, *available at* http://www.ftc.gov/os/comments/mobilecramming/564482-00017-86205.pdf.

[93] *See* Wayne Flower, *False Text Message Racket Cashes in on Mobile Phone Charges*, HERALD SUN, June 23, 2013, *available at* http://www.heraldsun.com.au/news/law-order/false-text-message-racket-cashes-in-on-mobile-phone-charges/story-fni0fee2-1226668456799; Katia Moskvitch, *Hidden Mobile Charges that Could Be Buried in Your Bill*, BBC NEWS, Sept. 6, 2012, *available at* http://www.bbc.co.uk/news/technology-19402398.

Address Online and Mobile Threats."[94] In it, they state that "'cramming' to Premium Rate 'love advice' or other text message services by Affiliates and/or Content providers has been commonplace."[95]

Australia has reported some success in combatting mobile cramming. In a case study published by the Australian Competition & Consumer Commission ("ACCC"), the ACCC found that previous controls – including industry self-regulation and some investigation into specific merchants – were not effective in reducing cramming. To address the problem, Australian authorities increased consumer education, brought enforcement actions, and worked closely with industry on adopting a set of regulations to protect consumers, which the ACCC noted was a "crucial component of providing a long-term solution to the problem."[96] Those regulations included, for example, requiring that carriers offer a block on Premium SMS services to consumers.[97] After taking these steps, Australian authorities reported that the number of consumer complaints in Australia about cramming have dropped significantly.[98]

IV. STRATEGIES TO REDUCE MOBILE CRAMMING

Stakeholders in the mobile billing field have adopted a number of strategies to attempt to address unauthorized third-party charges. Below we discuss the current strategies being used to combat mobile cramming in four areas, and provide FTC staff's view on further action to be taken to address the problem. Specifically, industry participants should give consumers the option of blocking third-party charges, implement measures to detect and prevent crammed charges from appearing on mobile phone bills, provide adequate disclosures of third-party charges, and establish clear and consistent dispute resolution policies for unauthorized charges. Staff's recommendations apply regardless of what form of carrier billing is being used, whether it is DCB, Premium SMS, or a new form of carrier billing that emerges.

---

[94] LONDON ACTION PLAN & M³AAWG, BEST PRACTICES TO ADDRESS ONLINE AND MOBILE THREATS (Oct. 15, 2012), *available at* http://www.maawg.org/sites/maawg/files/news/M3AAWG_LAP_Best_Practices_to_Address_Online_and_Mobile_Threats_0.pdf.

[95] *Id.* at 36.

[96] AUSTRALIAN COMPETITION & CONSUMER COMM'N, MOBILE PREMIUM SERVICES: MEETING THE CHALLENGES (Sept. 18, 2012), at 1, 5-8, *available at* http://www.accc.gov.au/system/files/Mobile%20Premium%20Services%20-%20Case%20Study.pdf.

[97] *Id.* at 8.

[98] *Id.* at Figure 2.

A. Allowing Consumers to Avoid Third-Party Charges to Mobile Accounts

Consumers should be able to choose to avoid third-party charges to their mobile accounts altogether and, as explained below, some carriers appear to have taken steps to allow consumers to do so. FTC staff reiterates prior Commission recommendations in this area and recommends further industry best practices to empower consumers to make informed choices about third-party charges.

1. Current Industry Practices

According to CTIA, the four major carriers currently permit their customers to block third-party charges from being placed on their phone accounts.[99] This option gives consumers who may not wish to use their phone accounts as a payment mechanism the ability to avoid third-party charges altogether, including unauthorized charges.[100] Although some carriers provide information about these blocking options in their service agreements, at the point of sale, and on their consumer-facing websites, several stakeholders have suggested that customer service representatives are poorly informed about the option, and that mechanisms to add or remove a block have not always been prominently featured or clearly described in the service agreement or on carrier websites.[101] Thus, some consumer groups have recommended prohibiting all third-party charges on wireless accounts by default, with an exemption for charitable giving.[102]

---

[99] *See* FTC Roundtable transcript, M. Altschul, at 34-35, 171-72, 177; Verizon Wireless Comments, *supra* note 6, at 4; Sprint Nextel Comments, *supra* note 6, at 4.

[100] Some carriers also provide the ability to block charges from a specific third party's services. *See* FTC Roundtable transcript, M. Altschul, at 172, 177.

[101] *See id.*, at 34-35; Verizon Wireless Comments, *supra* note 6, at 5; Sprint Nextel Comments, *supra* note 6, at 4; FTC Roundtable transcript, C. Witteman, at 92; California Public Utilities Commission Staff, Effectiveness of the Cramming Rules in Decision 10-10-034 in Protecting California Consumers from Unauthorized Charges on Their Phone Bills, and Related Developments in the Wireless Industry 16-17 (2014), *available at* http://www.cpuc.ca.gov/PUC/hottopics/2Telco/01_June_2014_Cramming_Report.htm; *see also* Nat'l Consumers League, et al., *Comments for the FTC Mobile Cramming Roundtable* (June 7, 2013) at 8, *available at* http://www.ftc.gov/sites/default/files/documents/public_comments/2013/06/564482-00013-86076.pdf [hereinafter "Consumer Groups FTC Comment"]. Additionally, the State of California requires phone companies to disclose to consumers that they can block all third-party charges on their mobile bills. *See* FTC Roundtable transcript, C. Witteman, at 90, 92; CPUC Rule, *supra* note 57, Attach. A at 4. From the record, it is not clear whether all carriers currently have this option and make these disclosures.

[102] *See* Comments of Center for Media Justice, Consumer Action, Consumer Federation of America, Consumers Union, National Consumer Law Center – on Behalf of Its Low-Income Clients, and National Consumer League, FCC CG Docket No. 11-116, at 18-19 (June 25, 2012), *available at* http://apps.fcc.gov/ecfs/document/view?id=7021977710 [hereinafter "Consumer Groups FCC Comment"]. The FCC is considering whether to modify its Truth-in-Billing rules to include a default block on all third-party charges. *See* FCC Refresh the Record, *supra* note 42. This would mean that consumers would have to affirmatively opt-in to third-party charges before such charges could be placed on their mobile phone bills.

During the mobile cramming roundtable, industry participants stated that some third-party charges originate on devices used by children.[103] When children incur charges on a family plan, the primary account holder – typically a parent or other adult – receives a single monthly bill that includes charges for all the family members' mobile devices that share the account. According to CTIA, mobile carriers offer account holders the opportunity to block all third-party charges on a family account or to block individual mobile devices from adding charges to the account.[104] As discussed above in Part III, however, evidence shows that many consumers do not understand that mobile telephone bills may contain third-party charges.

2.   FTC Staff Recommendations

As the Commission stated in comments to the FCC, all wireless providers should give consumers the option to block all third-party charges from their mobile phone accounts.[105] Providing a blocking option would significantly benefit consumers who wish to avoid third-party charges while imposing minimal costs to consumers who wish to use their mobile accounts for third-party billing.[106] At activation, consumers should be informed that third-party charges may be placed on their accounts, including using any phone numbers on a family plan, and they should be given the opportunity to block all charges at that time. This option should be clearly and prominently disclosed to consumers while the accounts are active, including on the carriers' websites. According to CTIA, carriers currently offer the ability to block third-party charges on specific phone numbers on a family plan.[107] Given the concerns raised about cramming on children's numbers, carriers should clearly and prominently disclose these options as well. Notably, the FCC's current Truth-in-Billing rules require landline carriers that offer blocking of third-party charges to clearly and conspicuously notify consumers of this option on each bill, their websites, and at the point of sale.[108]

---

[103] FTC Roundtable transcript, K. McCabe, at 61; A. Sege, at 130.

[104] *See id.*, M. Altschul, at 171, 177.

[105] *See* FTC Reply Comment, *supra* note 4, at 12.

[106] In earlier comments to the FCC, the Commission supported a ban or default block on third-party billing for *landline* billing, but stated that additional information was needed to determine whether a default block was appropriate in the mobile context. *See* FTC Reply Comment, *supra* note 4, at 3, 11-12. The Commission noted that, in contrast to landline third-party billing, the mobile billing platform has been used for some legitimate purposes – including charitable giving – and it is a potential platform for consumers to fund mobile payments for desired services. Indeed, some consumer advocacy groups have agreed that there is legitimate commerce occurring through mobile billing. *See* Consumer Groups FTC Comment, *supra* note 101, at 9. *See also* FTC Roundtable transcript, D. Derakhshani, Consumers Union, at 174. Based on the current evidence regarding authorized third-party carrier billing, as well as the potential costs and benefits of blocking third-party charges by default, staff does not believe that a default block on mobile third-party billing is warranted at this time.

[107] *See* FTC Roundtable transcript, M. Altschul, at 171-72, 177.

[108] *See* Report and Order and Further Notice of Proposed Rulemaking, Docket No. 11-116, at 59 (Apr. 27, 2012), *available at* http://hraunfoss.fcc.gov/edocs_public/attachmatch/FCC-12-42A1.pdf [hereinafter "FCC Report and Order"]; 47 C.F.R. § 64.2401(f).

Additionally, carriers should consider offering consumers the ability to block or allow only specific providers, or to block commercial providers only. In the context of Premium SMS, some carriers provided the option to block individual short codes or subscriptions.[109] Allowing more granular blocking would allow consumers to continue to authorize some third-party charges, including charitable or political donations, and may significantly benefit consumers who wish to use their mobile accounts for only certain kinds of third-party charges.[110]

B.   Strategies to Detect and Prevent Mobile Cramming

Industry participants have adopted a range of strategies to attempt to detect and prevent mobile cramming. This section focuses on strategies involving two key issues: avoiding deceptive practices that lead to unauthorized charges on mobile accounts, and ensuring that consumers are providing express, informed consent to third-party charges on mobile accounts. As explained below, industry efforts have fallen short on both fronts. For example, carriers are not taking sufficient action against merchants with high refund rates. Further, the voluntary standards developed by industry have largely focused on text-message based Premium SMS services and have not specifically addressed other types of carrier billing including app- or mobile web-based billing using DCB arrangements. Accordingly, staff recommends best practices for improvement in this area.

1.   Avoiding Deceptive Practices That Lead to Unauthorized Mobile Charges

a.   Industry Practices

Stakeholders in the mobile billing industry generally have relied on a set of voluntary guidelines to prevent deceptive marketing or advertising of mobile-billed goods or services. As discussed below, various industry participants monitor merchants' marketing and advertising practices, and track merchants who have previously engaged in deceptive behavior, to varying degrees. Despite these industry best practices, however, there have been significant problems with enforcement and compliance. Moreover, it is unclear how these best practices – many of which were developed in connection with the text-message based Premium SMS model – are being applied in the context of DCB arrangements.

---

[109] *See* FTC Roundtable transcript, M. Altschul, at 171-72, 177.

[110] Carriers should ensure that their blocking services provide the protections they are purported to provide. The Senate Commerce Committee's report on landline cramming found that consumers often incurred additional crammed charges on their landline telephone bills even after they requested the blocking of third-party charges on their landline telephone accounts. *See* S. COMMERCE COMM. CRAMMING REPORT, *supra* note 22, at 33-35, App. A. Mobile carriers should take steps to ensure that the blocking options they offer consumers are effective.

Voluntary guidelines. Until recently, the Mobile Marketing Association ("MMA"), a trade association that promotes mobile marketing, had taken the lead in publishing best practices for merchants who wish to place charges on mobile phone bills using Premium SMS.[111] In January 2013, CTIA began publishing a "Mobile Commerce Compliance Handbook," which aims to distill the principles of the MMA best practices to a shorter set of guidelines. The CTIA guidebook which describes itself as a "unified standard of compliance for mobile carrier billing," notes that its "requirements are based on the CTIA and participating carriers' experience with standard rate and premium shortcode [Premium SMS] programs," but that "the core rules are relevant to many types of mobile services."[112] It is not clear, however, how or to what extent the CTIA guidelines are used across all carrier billing arrangements.

The MMA's best practices include fairly detailed requirements for merchants to advertise their products and obtain consumers' authorization for third-party charges billed to a mobile phone account. Some commenters have noted, however, that merchants technically could comply with guidelines while hiding key information, such as the fact that entry of a mobile phone number to access purportedly free content will automatically subject the consumer to a recurring $9.99 charge on a monthly phone bill.[113] Specifically, a set of related companies that the State of Texas sued for cramming technically complied with an MMA guideline that a price disclosure be 125 pixels away from a box to enter a cell phone number on a website, but blended the price disclosure with a background image to make it difficult to see, as illustrated on the following page:[114]

---

[111] The MMA periodically revised and published these as the "U.S. Consumer Best Practices for Messaging." The most recent version (7.0) was published on October 16, 2012. *See* MOBILE MARKETING ASS'N, U.S. CONSUMER BEST PRACTICES FOR MESSAGING VERSION 7.0 (Oct. 16, 2012), *available at* http://www.mmaglobal.com/uploads/Consumer-Best-Practices.pdf. However, the MMA has stated that it will make no further revisions. *See* FTC Roundtable transcript, C. Frey, Mobile Marketing Association, at 76, 106-07.

[112] CTIA – THE WIRELESS ASSOCIATION, COMPLIANCE ASSURANCE SOLUTION: MOBILE COMMERCE COMPLIANCE HANDBOOK (November 4, 2013), *available at* http://www.wmcglobal.com/assets/ctia-mobile-commerce-compliance-handbook-v-1-3.pdf [hereinafter "CTIA Handbook"].

[113] *See, e.g.,* FTC Roundtable transcript, P. Singer, at 108-09.

[114] *See id.,* at 85-86; FTC Roundtable, *PowerPoint Presentation re: State of Texas v. Eye Level Holdings*, at 6, PowerPoint slides *available at* http://www.ftc.gov/sites/default/files/documents/public_events/mobile-cramming-roundtable/state-texas-v-eye-level-holdings-et-al.ppt [hereinafter "Eye Level Holdings Presentation"].

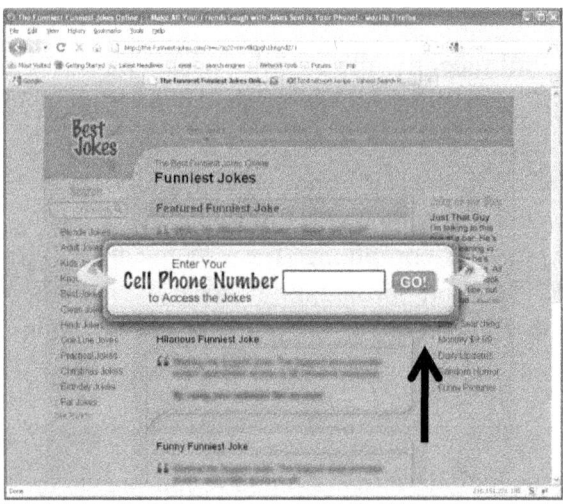

(vertical arrow added.) The defendants sent confirmation texts to consumers that included large gaps of spacing in the text, so that the consumer was likely to see the PIN code for opt-in at the top or bottom of the message, but unlikely to see the price disclosure buried inconspicuously in the middle of the text message, which would be unlikely to be displayed when the message was opened, as shown below:[115]

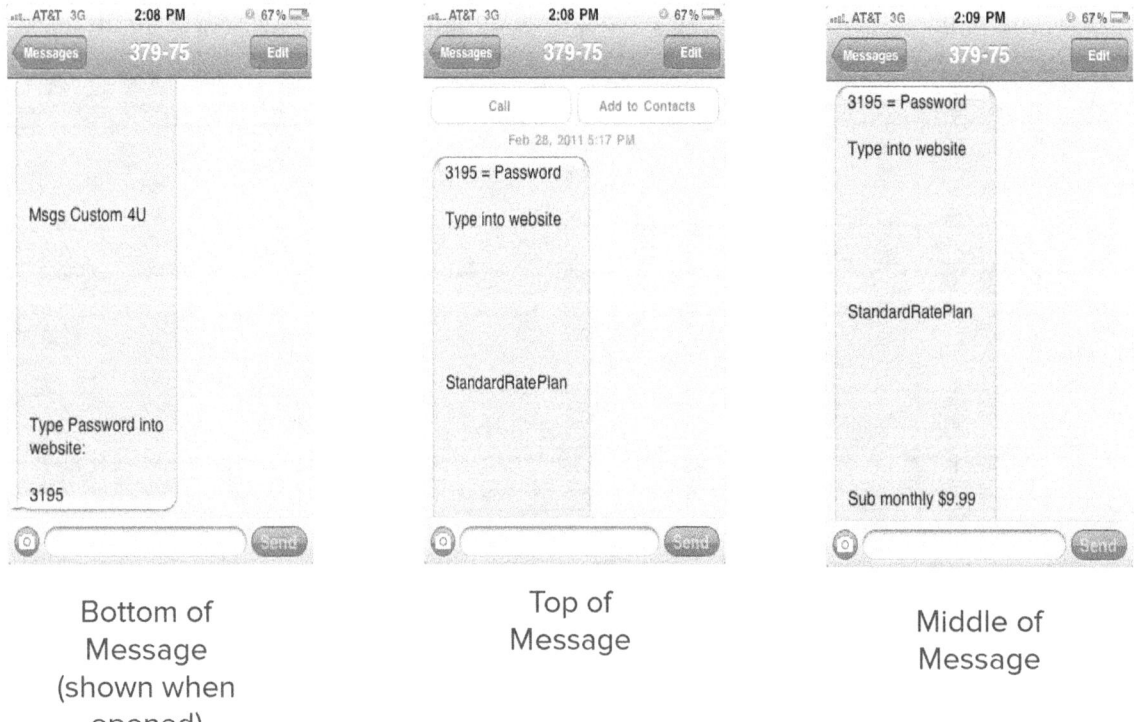

| Bottom of Message (shown when opened) | Top of Message | Middle of Message |

Monitoring and Enforcement. Although the MMA first assembled the set of best practices that became known as "guidelines" for the industry, it has no role in monitoring industry participants to

---

[115] *See* FTC Roundtable transcript, P. Singer, at 86; Eye Level Holdings Presentation, *supra* note 114, at 8.

enforce the guidelines.[116] Instead, industry participants choose whether to monitor and take action against merchants that violate the guidelines or otherwise engage in deceptive advertising or marketing practices.[117]

For example, carriers decide whether to allow merchants to bill on their networks, and prior to such approval, most carriers will review merchants' marketing materials and opt-in processes to ensure that they comply with the industry guidelines.[118] Each carrier then determines how much effort to invest in monitoring for compliance. At least some carriers have performed "in-market monitoring" – often by hiring third-party auditors – to check that the advertising and purchase flow continue to be compliant with industry or carrier-specific guidelines, at times looking beyond mere compliance with MMA and CTIA guidelines.[119] Further, at least one aggregator has noted that it also has engaged in in-market monitoring of campaigns.[120]

To be reasonably effective, this kind of monitoring must take into account challenges from determined fraudsters. For example, one entity used sophisticated cloaking software that displayed non-deceptive webpages to auditing companies, but displayed a different webpage to consumers.[121] Industry actors state that they have learned from experience in dealing with this kind of evasive activity.[122] Beyond reviewing and monitoring merchants' marketing practices for deceptive behavior, CTIA and some carriers also have vetted merchants up front based on their past practices – in short, attempting to identify potential bad actors in advance and prevent them from billing on carrier networks. While this monitoring of relationships can identify individuals or entities associated with fraudulent behavior in some contexts, it does not necessarily look closely at whether the merchant has been involved with unauthorized billing on landline phone bills, which some commenters suggest would be appropriate given the prevalence of cramming on landline phone bills.[123]

b.   Staff Recommendations

Under the FTC Act, merchants are in the first instance responsible for ensuring that their practices – including any advertising, marketing, and opt-in processes – are not deceptive, pursuant to the FTC Act. Further, information about price is important to consumers and should be disclosed clearly

---

[116] *See* FTC Roundtable transcript, C. Frey, at 76.

[117] *See, e.g.,* Wise Media TRO Memo, *supra* note 26, at 12-13.

[118] *See, e.g.,* Verizon Wireless Comments, *supra* note 6, at 2; Sprint Nextel Comments, *supra* note 6, at 5; FTC Roundtable transcript, M. Altschul, at 25-26, 44-45; *id.,* J. Bruner, at 81.

[119] *See id.,* J. Bruner, at 81, 96-97, 111; Verizon Wireless Comments, *supra* note 6, at 3-4; Sprint Nextel Comments, *supra* note 6, at 5.

[120] *See* FTC Roundtable transcript, A. Sege, at 141.

[121] *See id.,* P. Singer, at 87; *id.,* J. Bruner, at 97; Order and Preliminary Injunction, *Cellco P'ship v. Hope*, No. CV11-0432-PHX-DGC (D. Ariz. May 11, 2011), at 4-6.

[122] *See* FTC Roundtable transcript, J. Bruner, at 97-98.

[123] *See id.,* J. Bruner, at 104; *id.,* J. Breyault, at 13; Consumer Groups FTC Comment, *supra* note 101, at 9-10.

and conspicuously before charging a consumer's telephone account for a good or service. As FTC staff stated in its 2013 report, *.com Disclosures: How to Make Effective Disclosures in Digital Advertising*,

> Disclosures that are an integral part of a claim or inseparable from it should not be communicated through a hyperlink. Instead, they should be placed on the same page and immediately next to the claim, and be sufficiently prominent so that the claim and the disclosure are read at the same time, without referring the consumer somewhere else to obtain this important information. This is particularly true for cost information or certain health and safety disclosures.[124]

Thus, at a minimum, pricing information should be on the same page and immediately next to the purchase or buy button, entry of a PIN, or other invitation for a consumer to agree to a charge for a product or service, whether through a Premium SMS, DCB, or some other arrangement, and this price information should be prominent and in a font and size sufficient to read. Further, advertising and purchase confirmation screens must clearly disclose that the charge is being billed to a specific telephone account. Companies may consider, for example, whether identifying the last four digits of the consumer's phone number on a purchase confirmation screen – similar to the way that the last four digits of a credit card number are commonly listed on confirmation screens – is feasible and useful as part of the key disclosure that the charge will be placed on a telephone account.[125] While industry guidelines have in the past focused extensively on the text-message based Premium SMS opt-in process, the basic consumer protection principles outlined here should apply regardless of the type of carrier billing used, whether Premium SMS or DCB.

Staff also recommends that carriers and billing intermediaries implement reasonable procedures to scrutinize risky or suspicious merchants and terminate or take other appropriate steps against companies engaging in unlawful practices. These kinds of measures are present in other payment industries. In the credit card and ACH industries, for example, banks that provide payment processing services are subject to KYC ("Know Your Customer") procedures, which require banks to perform due diligence on their merchant-customers. Among other things, KYC principles require banks to gain familiarity with a merchant's marketing practices to identify high-risk activities and ensure that it is not

---

[124] *See* FED. TRADE COMM'N STAFF, REVISED .COM DISCLOSURES: HOW TO MAKE EFFECTIVE DISCLOSURES IN DIGITAL ADVERTISING 10 (2013), *available at* http://www.ftc.gov/sites/default/files/attachments/press-releases/ftc-staff-revises-online-advertising-disclosure-guidelines/130312dotcomdisclosures.pdf.

[125] Although the MMA and CTIA guidelines already purport to require pricing and payment information, the examples cited in the guidelines do not always make clear that it is the consumer's phone bill that will be charged. *See* MOBILE MARKETING ASS'N, *supra* note 111, at 23 (Best Practice 2.3-5) ("All advertising must clearly disclose the subscription term, billing interval and information on how the charges will be applied (*i.e.*, that the charges will be billed on the customer's wireless phone bill or deducted from the customer's prepaid balance)."); *but compare* CTIA Handbook, *supra* note 112, at 19-20, *and* MOBILE MARKETING ASS'N, *supra* note 111, at 38, 40. Ensuring that the consumer understands the billing mechanism is consistent with rules adopted by the Commission in other contexts. *See* FTC Telemarketing Sales Rule, 16 C.F.R. § 310.4(a)(7)(i)-(ii).

facilitating fraudulent or other illegal activity.[126] Further, the operating rules established by the credit card associations and NACHA – The Electronic Payments Association (for ACH transactions) encourage banks to monitor merchant transaction activity to identify merchants engaged in fraud or other illegal conduct, and to cut off their access to the payment systems if appropriate.[127] While the specific procedures may vary by industry, this kind of monitoring and termination of fraudulent merchants has been beneficial in protecting consumers and maintaining the integrity of the billing platform.

Similarly, staff recommends that if a carrier or billing intermediary discovers that a merchant has run a campaign containing deceptive advertising, or discovers the merchant engaged in unauthorized billing on landline phones, the carrier should closely monitor other campaigns run by that third party or its affiliates to ensure compliance. Carriers can use monitoring techniques to address known tactics by fraudsters to evade detection, in order to determine consumers' actual experience with third-party merchants' advertisements and sign-up processes, such as viewing advertisements on devices not registered to the carrier. Industry participants also can adopt a policy of terminating serious and repeat offenders. While there are costs to effective monitoring, there are also benefits both to the industry itself, in the form of lower costs to process refund requests and handle customer complaints, and to consumers, who avoid being crammed with unauthorized charges.

2.   Obtaining Consumers' Express, Informed Consent

a.   Current Industry Practices

As discussed above, mobile industry participants created the MMA best practices and the CTIA guidelines in an effort to ensure that consumers authorize the third-party charges that mobile carriers include on mobile telephone accounts. In the context of Premium SMS, the core of those self-regulatory initiatives is a requirement that third-party merchants employ a "double opt-in" process that requires consumers to take two separate steps to authorize a third-party charge, one of which is intended to confirm that the consumer is in physical possession of a phone tied to the account that will be

---

[126] *See, e.g.*, Fed. Deposit Ins. Corp., FDIC Supervisory Approach to Payment Processing Relationships With Merchant Customers That Engage in Higher-Risk Activities, FIL-43-2013 (Sept. 27, 2013*), available at* http://www.fdic.gov/news/news/financial/2013/fil13043.pdf; Fed. Deposit Ins. Corp., Payment Processor Relationships Revised Guidance, FIL-3-2012 (January 31, 2012), *available at* http://www.fdic.gov/news/news/financial/2012/fil12003.pdf; Office of the Comptroller of the Currency, OCC Bulletin 2013-29 (Oct. 30, 2013), *available at* http://occ.gov/news-issuances/bulletins/2013/bulletin-2013-29.html; Office of the Comptroller of the Currency, OCC Bulletin 2006-39 (Sept. 1, 2006), *available at* http://www.occ.gov/news-issuances/bulletins/2006/bulletin-2006-39.html. Banks also require their payment processor customers to comply with KYC procedures for merchant-customers.

[127] *See, e.g.*, VISA, ACQUIRER BEST PRACTICES CONTINUITY MERCHANTS (Aug. 26, 2010), *available at* https://usa.visa.com/download/merchants/bulletin_acquirer_continuity_best_practices.pdf; NACHA Operating Rules, *supra* note 61, at § 2.2.3; *see also* VISA, VISA E-COMMERCE MERCHANTS' GUIDE TO RISK MANAGEMENT 24-25 (2008), *available at* https://usa.visa.com/download/merchants/visa-risk-management-guide-ecommerce.pdf.

charged.[128] As noted above, it is not clear whether industry participants in DCB arrangements routinely follow the same sort of double opt-in process used for Premium SMS transactions or use other methods to obtain consumers' authorization.

Individual carriers have also used refund rates to identify third-party merchants who may not be obtaining consumer consent to charges, and individual carriers have the discretion to penalize or cut off merchants with high refund rates,[129] with one carrier, for example, terminating Premium SMS merchants with an 8% or higher refund rate.[130] Based on the record of consumer complaints, refunds, and law enforcement actions described above, however, it does not appear that the double opt-in requirement or carriers' current practices involving monitoring refund rates and terminating problematic merchants have prevented unauthorized charges on consumers' mobile telephone accounts. Commenters have noted that crammers can easily obtain consumers' phone numbers and find ways to bill consumers even without completing either step of the purported double opt-in.[131] Consumers have repeatedly complained that charges appear on accounts without *any* affirmative action by the consumer.[132]

Given the issues with spurious consumer authorizations, one question raised by various stakeholders is which entity in the carrier billing process should host the records and control the process of the consumer authorization. Although merchants historically have controlled the opt-in process for Premium SMS,[133] there appears to be a trend toward greater carrier or aggregator control of the opt-in process.[134] With respect to noncommercial third-party charges, one representative of the Mobile Giving

---

[128] *See* FTC Roundtable transcript, A. Sege, at 128; *id.*, J. Bruner, at 113. Common forms of double opt-in include:

1. Mobile phone opt-in – a consumer sends a text message with a keyword to a short code, receives a confirmation text, and then sends a second text message to confirm the transaction;
2. Web opt-in – a consumer first enters his or her number on a website, then receives a text with a PIN, which the consumers enters back into the website;
3. WAP (wireless access protocol) – a consumer responds to an offer on his or her phone (displayed through the mobile web), and clicks on a confirmation button on the phone two separate times; and
4. IVR (interactive voice response) – a consumer opts-in over the phone through a toll-free number and confirms the transaction twice. *See* FTC Roundtable transcript, J. Sizer, Aegis Mobile, at 122-23.

[129] *See* Wise Media TRO Memo, *supra* note 26, at 12-13; FTC Roundtable transcript, J. Bruner, at 82; *id.*, M. Altschul, at 146.

[130] *See* Wise Media Declaration, *supra* note 20, at FTC 1 - 000012-13, ¶ 27.

[131] *See* FTC Roundtable transcript, M. Tiano, S. Comm. on Commerce, Science, and Transp., at 164; NAAG Comments, *supra* note 47, at 11-12; Consumer Groups FCC Comment, *supra* note 102, at 10-15.

[132] *See supra* Part III.A, at 11, 14.

[133] *See* FTC Roundtable transcript, M. Altschul, at 39, 62; *id.*, J. Manis, at 27.

[134] *See id.*, M. Altschul, at 39, 62; Verizon Wireless Comments, *supra* note 6, at 3. At least one major aggregator indicated that it will require all merchants to use the aggregator to host the opt-in records. *See* FTC Roundtable transcript, A. Sege, at 167-68. The State of Texas, however, recently sued that aggregator for allegedly "orchestrating and facilitating" a cramming scheme. *See* Mobile Messenger Complaint, *supra* note 54, at 5.

Foundation, a billing intermediary, noted that it manages the text-based opt-ins for participating charities.[135] And at least some DCB intermediaries also appear to manage the consumer opt-in, including by sending the consumer a PIN to confirm authorization.[136] Centralization allows for greater control and transparency as to individual merchants' authorization and billing activities.[137] One direct carrier billing intermediary has stated that managing the two-step confirmation process is one of the factors that has led to its low refund rate, reportedly less than 1% on some carriers.[138]

As mobile cramming issues have become more apparent, state regulators and other stakeholders have called for greater centralization of opt-in records as one way to address cramming.[139] The Attorneys General from thirty-six states, the District of Columbia, Guam, Puerto Rico, and the Virgin Islands have criticized current industry practices for placing too much of the opt-in responsibility in the hands of third-party merchants and billing intermediaries. In the view of those Attorneys General, those parties have significant incentive to bill for services whether or not consumers have authorized them.[140]

An order by the CPUC in 2010 sets forth the general principle that "the billing telephone corporation [is the entity that] bears ultimate responsibility for all items presented in a subscriber's bill."[141] In 2013, the National Association of State Utility Consumer Advocates (NASUCA) recommended the passage of state legislation to address mobile cramming.[142] The proposed legislation would hold telephone companies, aggregators, and third-party merchants all responsible for unauthorized charges, providing a defense only if "the circumstances giving rise to the unauthorized charge were beyond the control of, or could not reasonably have been prevented by, the company."[143]

b. Staff Recommendations

It is critical that consumers provide their express, informed consent to charges before they are billed to a mobile account, and that reliable records of such authorizations are maintained. Industry practices have not sufficiently prevented unauthorized charges. In Premium SMS, mobile carriers typically have relied on the merchant's representation – passed on by the billing intermediary – that a

---

[135] *See* FTC Roundtable transcript, J. Manis, at 27.

[136] *See id.*, M. Niejadlik, at 136; *id.*, J. Greenwell, at 65.

[137] *See* FTC Roundtable transcript, J. Manis, at 26-27.

[138] *See id.*, J. Greenwell, at 65.

[139] *See* FTC Roundtable transcript, C. Witteman, at 114-15.

[140] *See* NAAG Comments, *supra* note 47, at 12.

[141] CPUC Rule, *supra* note 57, Attach. A at 4; *see also* FTC Roundtable transcript, C. Witteman, at 90.

[142] Nat'l Ass'n of State Util. Consumer Advocates, *Urging State Legislatures to Prohibit the "Cramming" of Unauthorized Charges Onto Consumer Telephone Bills* (June 11, 2013), at 2, *available at* http://nasuca.org/2013-01-urging-state-legislatures-to-prohibit-the-cramming-of-unauthorized-charges-onto-consumer-telephone-bills-and-proposing-a-statute-to-solve-the-problem/.

[143] *Id.* at 7.

consumer opted-in to a charge, but, as the law enforcement actions and consumer complaints described above demonstrate, those representations often are unreliable.

Given the unreliability of many merchants' claims that they have obtained consumer consent, more centralized control of the consumer opt-in process and authorization records is needed. As noted above, the industry appears to have already implemented changes in this direction. As the technology used for consent management shifts from Premium SMS arrangements to DCB arrangements, carriers and intermediaries should maintain sufficient control over the consent process to address the problem of unauthorized charges, for example, by maintaining reliable and accessible records of consumer authorizations. Centralization may shift some compliance costs from the merchants to carriers and intermediaries, but should benefit consumers and industry participants by making it more difficult for unscrupulous merchants to place unauthorized charges and by streamlining dispute resolution when a consumer claims a charge was unauthorized.

Further, individual carriers' policies on taking action against merchants with high refund rates have not been sufficient to combat mobile cramming – for example, as noted above, at least one carrier continued billing consumers for Wise Media charges even after the company's monthly refund rate exceeded 40% on some campaigns.[144] Mobile carriers should implement policies, or strengthen existing polices, to investigate and take appropriate action when consumer refund requests and complaints indicate that a merchant may be cramming charges without consumers' consent. As discussed above, other industries have examined refund rates to determine what rates should trigger additional scrutiny and possible termination of a merchant.[145] Monitoring consumer refund requests, and taking appropriate action when there are indications of unauthorized charges, can be a highly effective means of detecting and stopping cramming. It is particularly important to monitor refund requests in carrier billing, where merchants have been able to cram charges without using deceptive advertisements or other techniques to contact consumers that can be independently identified.

C. Adequate Disclosures of Third-Party Charges

Another important step in preventing cramming is ensuring that consumers are adequately informed of all third-party charges on their accounts. Many consumers currently overlook third-party charges on their mobile accounts. Staff provides recommendations in this section for making disclosures of those charges clearer and more prominent.

---

[144] *See* Wise Media Declaration, *supra* note 20, at Ex. FTC 21.

[145] *See supra* Part III.B.1, at 16.

1. Current Industry Practices

As discussed above in Part III, many consumers have found it difficult to detect unauthorized charges on their mobile phone bills because of the description and location of third-party charges on the bill.[146] For example, the State of Texas alleged in one case that crammed charges were sometimes listed as "Standard Rate Plan" on the consumers' bills.[147] In the FTC's case against Wise Media, the charges typically appeared buried among other charges; as noted above, one consumer reported that a Wise Media charge was listed on page 18 of her phone bill. Wise Media charges also appeared in a manner that made it difficult, if not impossible, to identify that the charges were for services provided by a company other than the carrier.[148] And in its complaint against T-Mobile, the Commission has alleged that consumers who reviewed an online summary of their bills would see third-party charges lumped together in a line item labeled "Use Charges" that could include charges for both T-Mobile's own services, such as for text messages, and for third-party charges.[149] Roundtable panelists noted that many consumers are simply unaware that third parties can even place charges on mobile phone bills, and thus are unlikely to review complicated and lengthy phone bills to find such charges and make sure they are authorized.[150]

The FCC's current Truth-in-Billing rules require mobile carriers' bills to clearly identify the service provider and provide a full and non-misleading description of each charge.[151] In the landline space, there are additional FCC requirements that carriers place all third-party charges for non-telecommunications services in a distinct bill section separate from all carrier charges, and that the bill must provide separate totals for carrier and non-carrier charges.[152] Although these latter requirements currently apply to landline bills only, the FCC recently reopened its request for comment on whether to expand them to mobile phone bills.[153]

2. Staff Recommendations

To help consumers understand what third-party services they are paying for with their mobile phone accounts – and determine whether there are any unauthorized charges on their accounts – carriers should clearly and conspicuously disclose all charges for third-party services in a non-deceptive manner.

---

[146] *See supra* Part III, at 10, 20.

[147] *See* FTC Roundtable transcript, P. Singer, at 86-87.

[148] *See, e.g.*, Wise Media Declaration, supra note 20, at Exs. FTC 66, FTC 71, FTC 72, FTC 73; Wise Media TRO Memo, *supra* note 26, at 6 n.2.

[149] T-Mobile Complaint, *supra* note 20, at ¶ 13.

[150] *See* FTC Roundtable transcript, P. Singer, at 118; *id.*, M. Niejadlik, at 155-56; *id.*, K. McCabe, at 12, 28.

[151] *See* FTC Roundtable transcript, L. Follansbee, Fed. Commc'ns Comm'n, at 133; FCC Report and Order, *supra* note 108, at 6; 47 C.F.R. § 64.2401(b).

[152] *See* 47 C.F.R. § 64.2401(a); FCC Report and Order, *supra* note 108, at 20.

[153] *See* FTC Roundtable transcript, L. Follansbee, at 133-34; FCC Report and Order, *supra* note 108, at 20-21, 53; FCC Refresh the Record, *supra* note 42, at 1.

In particular, the name of the third-party service and any associated bill heading should relate to the product offered and not suggest an affiliation with the carrier's service. Billing intermediaries and merchants must provide accurate information to carriers for these disclosures.

FTC staff further recommends that carriers, as well as the FCC, consider ways to make third-party charges more conspicuous. For example consistent with the FCC's rules for landline billing, third-party charges could be placed in a separate part of consumers' mobile phone bills and separate subtotals for carrier and third-party charges be provided wherever total charges are disclosed – typically on the first page of the bill. As discussed above, many consumers are unaware that third-party charges can be placed on their mobile phone bills, and consumers have expressed frustration that the charges are often buried in lengthy phone bills. The benefits of enhanced disclosures, which will help consumers identify unauthorized charges, would likely outweigh the costs of any necessary changes to the format of the mobile phone bill.

Staff notes that these recommendations may be less beneficial for consumers who auto-pay their bills and may be especially unlikely to review the charges. Carriers should thus consider whether these consumers would benefit from receiving a separate notification of third-party charges. Consumers with prepaid phone plans who do not typically receive phone bills should also receive such a notification from the carrier.

D. Consumer Dispute Protections and Refunds

Stakeholders have expressed widely divergent views regarding consumers' ability to obtain refunds for unauthorized third-party charges under current industry practices. As noted below, industry representatives state that the carriers have generous refund policies that enable consumers to obtain refunds promptly from their carriers when they believe they have been crammed. By contrast, consumer advocates and regulators state that it is difficult for consumers to obtain refunds, and they also note that refunds often are limited to one or two months' worth of charges even when consumers discover they incurred crammed charges for a longer time period. FTC staff reiterates the Commission's prior recommendation that mobile carriers should provide a clear and consistent process for customers to dispute suspicious charges on their mobile accounts and obtain reimbursement.[154] Staff provides further recommendations for industry in implementing that process.

1. Current Industry Practices

Two major carriers submitted comments stating that they encourage consumers to contact them to discuss third-party charges or billing issues.[155] Further, according to CTIA, mobile carriers have generous refund policies with respect to third-party charges on consumer's telephone bills.[156] In particular, CTIA touts carriers' "one-and-done" policies, which empower the carriers' customer service representatives to address a consumer's complaint about unauthorized third-party charges during the

---

[154] FTC Reply Comment, *supra* note 4, at 12.

consumer's initial call to dispute the charge.[157] Additionally, in the landline context, the United States Telecom Association has stated that non-payment of third-party charges will not result in phone service being suspended or terminated.[158] Carriers, however, have not stated that this protection applies to the mobile context.

Consumers' experiences attempting to get refunds vary and are not consistent with a "one-and-done" policy.[159] In lawsuits that alleged cramming, many consumers reported that they found it difficult, if not impossible, to obtain refunds from their carriers when they disputed third-party charges.[160] Some consumers reported that they were only able to obtain partial refunds rather than full refunds for the entire period during which they incurred unauthorized charges.[161] Law enforcement officials have also noted that consumers have inconsistent experiences with obtaining refunds when they discover that they have been crammed.[162] Consumers' experiences include carriers providing partial refunds, carriers providing no refunds and instead referring consumers to the third-party merchant, and third-party merchants providing partial or no refunds.[163] In addition to calling for clear and consistent dispute policies, consumer advocates also recommend that consumers be able to withhold payment on disputed charges with the knowledge that their phone service will not be cut off while the investigation is pending.[164] Although the landline third-party billing industry states that it has taken this step, one roundtable participant raised concerns that consumers are unaware that they can withhold payment pending a dispute of a third-party charge on a mobile account.[165]

It is also not clear whether, and to what extent, carriers proactively notify or provide refunds to consumers who have been billed by merchants that the carrier has terminated for engaging in cramming. According to CTIA, "in appropriate cases," at least some carriers will proactively provide retroactive refunds without waiting for consumers to complain.[166] The carriers have not, however, provided any data regarding the frequency with which they have issued retroactive refunds to all consumers who incurred charges from a third party that the carriers terminated for cramming.

---

[155] *See* Sprint Nextel Comments, *supra* note 6, at 6; Verizon Wireless Comments, *supra* note 6, at 5.

[156] *See* FTC Roundtable transcript, M. Altschul, at 60.

[157] *Id.*, M. Altschul, at 185-86.

[158] FCC Workshop, *Bill Shock and Cramming* (Apr. 17, 2013), G. Reynolds, United States Telecom Association, at 99:15, *recording available at* http://www.fcc.gov/events/workshop-bill-shock-and-cramming.

[159] *See* FTC Roundtable transcript., M. Tiano, at 186-87 (discussing wireline cramming); *id.*, J. Breyault, at 55; *id.*, P. Singer 101-02, 116-17; Consumer Groups FTC Comment, *supra* note 101, at 7.

[160] *See* FTC Roundtable transcript, P. Singer, at 116-117; Wise Media TRO Memo, *supra* note 26, at 11-12.

[161] *See* FTC Roundtable, K. McCabe, at 61; KOLODINSKY, *supra* note 66, App. C at 6-8.

[162] *See* NAAG Comments, *supra* note 47, at 3-4.

[163] *See id.*

[164] FTC Roundtable transcript, J. Chilsen, at 121; Consumer Groups FCC Comment, *supra* note 102, at 21.

[165] FTC Roundtable transcript, J. Chilsen, at 121.

[166] *Id.*, M. Altschul, at 152.

2.   Staff Recommendations

As the Commission has previously emphasized, carriers should implement a clear and consistent process for consumers to dispute suspicious charges on their mobile accounts and obtain refunds for unauthorized charges.[167]   In the FTC Mobile Payments Report, FTC staff noted that no federal statutory protections have been applied to consumer disputes about unauthorized charges placed on mobile carrier accounts, in contrast to the implementation of statutory dispute rights for unauthorized credit card and debit card charges.[168]   For example, in the credit card industry, consumers have dispute resolution rights and liability limits for unauthorized charges under Regulation Z, including a right to withhold payment while the dispute is pending.[169]   For debit card users, Regulation E provides protection for consumers, including a requirement that funds debited in an unauthorized transaction be returned to a consumer's account within ten days, pending further investigation.[170]

Because consumers currently do not have access to a clear and consistent process for disputing unauthorized charges to their mobile accounts, staff recommends that carriers implement such processes and ensure that their customer service representatives abide by them. For example, as noted above, in the landline context, industry members have stated that consumers can withhold payment on disputed charges during the dispute period without a cut-off in phone service or accrual of interest. Consumers appear to be unaware of an analogous right in the mobile billing context, however. Staff recommends that this protection be extended to the mobile billing context, and that consumers be informed of it. Further, given the extensive evidence that consumers are unaware of third-party charges on their phone accounts, FTC staff recommends that when consumers seek refunds for recurring unauthorized charges and a carrier concludes those charges were crammed, consumers could be granted refunds for the same recurring charge in previous months to the extent it is practicable to identify those prior charges.

Finally, when a third party's billing activities are terminated for unauthorized charges, staff recommends that the carrier notify consumers who incurred charges from the third party to allow them to request a refund.[171]   Moreover, given the major carriers' recent decision to discontinue Premium

---

[167] FTC Reply Comment, *supra* note 4, at 12.

[168] FTC MOBILE PAYMENTS REPORT, *supra* note 6, at 5-8.

[169] *See* 12 C.F.R. §§ 1026.12, 1026.13.

[170] *See* 12 C.F.R. §§ 1005.6, 1005.11.

[171] The Senate Commerce Committee's report on landline cramming discusses a customer survey that a landline carrier conducted via email when it suspected that a third party had engaged in landline cramming. None of the consumers who responded to the survey reported that they had authorized the third-party charge at issue. None of the first twelve respondents to the survey had complained to the landline carrier about the unauthorized third-party charge before the landline carrier conducted the survey, and it is not known whether any of the other survey respondents had complained to the carrier before receiving the survey either. S. COMMERCE COMM. CRAMMING REPORT, *supra* note 22, at 37.

SMS services in response to documented allegations of fraud,[172] carriers should consider proactively notifying consumers who have paid a Premium SMS charge for merchants that the carriers have reason to believe are crammers and informing those consumers of the steps they should take to seek a refund if they believe the charge was unauthorized. In these circumstances involving clear fraud, consumers should have an opportunity to be made whole.

CONCLUSION

As consumers increasingly turn to their mobile phones as a payment mechanism, it is critical that carriers and other industry participants proactively address mobile cramming. Although the industry has taken some steps to combat unauthorized third-party charges on mobile phone accounts, more needs to be done to address this ongoing problem. Companies must keep basic consumer protection principles in mind as they develop and promote carrier billing options, particularly as the industry continues its transition from Premium SMS to other billing platforms like DCB. Mobile billing offers a promising option for consumers, but industry participants must recognize that innovation goes hand-in-hand with longstanding consumer protection principles. The more companies keep these principles in mind, the more consumers will trust and adopt these innovative payment methods.

The FTC will continue to monitor and, where appropriate, investigate industry participants – carriers, billing intermediaries, and merchants – involved in third-party mobile billing and bring further enforcement actions. Further, the FTC will continue to monitor the issue of cramming on mobile phone accounts and evaluate whether other potential solutions – including legislative measures and additional regulatory changes – are necessary to ensure consumers are protected from unwanted and unauthorized charges.

---

[172] *See T-Mobile Will No Longer Allow Third Parties to Bill Customers for Premium SMS Services*, T-MOBILE, http://newsroom.t-mobile.com/news/t-mobile-announces-new-program-related-to-premium-sms-charges.htm (last visited July 23, 2014) ("Despite protections and processes put in place by T-Mobile and the industry, not all Premium SMS vendors have acted responsibly."); Pamela Prah, *'Cramming' phone scams targeted by state attorneys general*, USA TODAY, Dec. 19, 2013, *available at* http://www.usatoday.com/story/news/nation/2013/12/19/stateline-cellphone-cramming/4129561/ (Verizon Wireless statement that it was in the process of winding down its premium messaging business, in part, because of "recent allegations that third parties have engaged in improper conduct in providing premium messaging services to our customers.").